ADVANCES IN
Applied Microbiology
VOLUME 32

ADVANCES IN

Applied Microbiology

Edited by ALLEN I. LASKIN

Somerset, New Jersey

VOLUME 32

 1987

ACADEMIC PRESS, INC.
Harcourt Brace Jovanovich, Publishers
Orlando San Diego New York Austin
Boston London Sydney Tokyo Toronto

COPYRIGHT © 1987 BY ACADEMIC PRESS, INC.
ALL RIGHTS RESERVED.
NO PART OF THIS PUBLICATION MAY BE REPRODUCED OR
TRANSMITTED IN ANY FORM OR BY ANY MEANS, ELECTRONIC
OR MECHANICAL, INCLUDING PHOTOCOPY, RECORDING, OR
ANY INFORMATION STORAGE AND RETRIEVAL SYSTEM, WITHOUT
PERMISSION IN WRITING FROM THE PUBLISHER.

ACADEMIC PRESS, INC.
Orlando, Florida 32887

United Kingdom Edition published by
ACADEMIC PRESS INC. (LONDON) LTD.
24–28 Oval Road, London NW1 7DX

LIBRARY OF CONGRESS CATALOG CARD NUMBER: 59-13823

ISBN 0–12–002632–5 (alk. paper)

PRINTED IN THE UNITED STATES OF AMERICA

87 88 89 90   9 8 7 6 5 4 3 2 1

CONTENTS

Microbial Corrosion of Metals

WARREN P. IVERSON

I.	Introduction	1
II.	Corrosion	4
III.	Microorganisms Involved in Corrosion	6
IV.	Mechanisms of Microbial Corrosion	11
V.	Microbial Corrosion of Ferrous Alloys	21
VI.	Microbial Corrosion of Nonferrous Metals and Alloys	22
VII.	Prevention and Control	27
VIII.	Conclusions	30
	References	31

Economics of the Bioconversion of Biomass to Methane and Other Vendable Products

RUDY J. WODZINSKI, ROBERT N. GENNARO, AND MICHAEL H. SCHOLLA

I.	Introduction	37
II.	Input Data	38
III.	Overview of Computer Programs	38
IV.	Substrates for Anaerobic Digestion	48
V.	Capital Costs	49
VI.	Process Conditions	49
VII.	Digester, Gas, Solids, and Liquid Effluent Process Designs	49
VIII.	Factors Affecting Profitability of Dairy Cattle Waste Digestion	53
IX.	Factors Affecting Profitability of Beef Feedlot Waste Digestion	57
X.	Factors Affecting Profitability of Swine Waste Digestion	64
XI.	Factors Affecting Profitability of Poultry Waste Digestion	64
XII.	Factors Affecting Profitability of Water Hyacinth Digestion	66
XIII.	Effects of Potential Process Improvements on the Economics of Methane Production	69
XIV.	Conclusions	84
	References	86

The Microbial Production of 2,3-Butanediol

ROBERT J. MAGEE AND NAIM KOSARIC

I.	Introduction	89
II.	Microbiology	90

III.	Biochemistry	95
IV.	Environmental Parameters	103
V.	Potential Substrates	124
VI.	Culture Techniques	132
VII.	Butanediol Recovery	135
VIII.	Process Design	139
IX.	Economics	152
X.	Conclusions	157
	References	158

Microbial Sucrose Phosphorylase: Fermentation Process, Properties, and Biotechnical Applications

ERICK J. VANDAMME, JAN VAN LOO, LIEVE MACHTELINCKX, AND ANDRE DE LAPORTE

I.	Introduction	163
II.	Producing Strains and Enzyme Assay	163
III.	Sucrose Phosphorylase Production	167
IV.	Properties of Sucrose Phosphorylase	172
V.	Kinetics of the Sucrose Phosphorylase Reaction	184
VI.	Immobilization of Sucrose Phosphorylase	188
VII.	Applications of Sucrose Phosphorylase	192
	References	199

Antitumor Anthracyclines Produced by *Streptomyces peucetius*

A. GREIN

I.	Introduction	203
II.	Cultural and Morphological Characteristics of S. peucetius	203
III.	Mutation and Selection	205
IV.	Biosynthesis	206
V.	Bioconversion	211
VI.	Fermentation	212
VII.	Conclusion	213
	References	213

INDEX 215

Microbial Corrosion of Metals

WARREN P. IVERSON*

Frederick, Maryland 21701

I. Introduction

Microbial corrosion may be defined for the purposes of this review as the deterioration or corrosion of metals induced by the activities of microorganisms. The field, therefore, by this definition is an interdisciplinary one; investigations in this field require an understanding of microbiology in addition to corrosion science. Consequently, for many years, the number of investigators in the world has remained quite small. A further hindrance has been a general lack of awareness of the significance of the problem. The organisms mainly implicated and primarily studied in relation to microbial corrosion are the sulfate-reducing bacteria (SRB). These organisms, in addition to being strict anaerobes, have been difficult to grow, isolate, and count. This has been an additional barrier to those who would even be interested in working in the field of microbial corrosion. Also, papers in this field have been widely scattered throughout the world in many obscure journals.

Within the last several years, the situation seems to be changing. The Second International Conference on Microbially Induced Corrosion was held at the National Bureau of Standards in 1985, the first being held in 1983 at the National Physical Laboratory, Teddington, England. The proceedings of a Workshop on Biodeterioration with emphasis on microbial corrosion, held in 1985 in Argentina have also been published (Argentine–U.S.A. Workshop, 1986). In addition, the National Association of Corrosion Engineers has, since 1983, continued to sponsor symposia in this field at their national meetings. Occasional papers on biological corrosion have been presented at the international symposia on biodeterioration and at the international congresses on metallic corrosion. Two cooperative groups, one in England and the other in the United States, continue to be active in microbial corrosion research. The microbial corrosion group in England, established in 1981, is sponsored by the International Biodeterioration Research Group, originally formed by The Organization for Economic Co-operation and Development (OECD) in 1962. A Biological Corrosion Committee was established in the United States by the National Association of Corrosion Engineers in 1981.

---

*Formerly with the National Bureau of Standards, Gaithersburg, Maryland 20899.

A number of bibliographies have been published. These include an out-of-date publication compiled by Lee (1963) and bibliographies by the Metals Society (1983) and the Biodeterioration Center (1981). A bibliography is now in preparation by Lutey (1985).

A number of booklets on microbiological corrosion have recently appeared in French (Chantereau, 1981), Portuguese (Videla, 1981), Swedish (Kucera, 1980), Spanish (Videla and Salvarezza, 1984), and in English (Pope et al., 1984). An older volume (Miller, 1970), primarily devoted to microbial corrosion, provides excellent background material. The National Association of Corrosion Engineers (1972) has published a small manual on biological corrosion in the oil field.

General reviews of microbial corrosion include those by Tiller (1982, 1983b), Miller (1981), Iverson (1974), and Costello (1969). Reviews devoted primarily to corrosion by SRB and sulfur-oxidizing bacteria (Cragnolino and Tuovinen, 1984) and by SRB (Hamilton, 1985) in the petroleum industry (Iverson and Olson, 1984) have appeared.

A. HISTORICAL PERSPECTIVE

Probably the first reported suggestion that microorganisms might be involved in metal corrosion was that by Garrett (1891). He attributed the corrosion of lead-sheathed cable to the action of bacterial metabolites. Gaines (1910) produced evidence that iron and sulfur bacteria were involved in the corrosion of the interior and exterior of water pipes by demonstrating the presence of abnormally large quantities of sulfur. Iron-depositing bacteria, *Caulobacter* and *Gallionella,* were later reported as being associated with deposits in pipes (Ellis, 1919; Harder, 1919).

Although reports of underground corrosion in the absence of stray electrical currents continued to accumulate, it was not until 1934 that von Wolzogen Kühr and van der Vlugt (1934) proposed their microbiological theory (cathodic depolarization) to account for the anaerobic corrosion of iron pipes. They proposed that this type of corrosion was caused by the obligate anaerobic SRB. Evidence for this bacterially associated corrosion continued to accumulate from around the world and was reviewed by Starkey and Wright (1945). (The mechanism of cathodic depolarization theory will be discussed in Section IV.) Since these early studies, there has been increasing evidence that other organisms, in addition to SRB, have been involved in the corrosion of iron as well as other metals.

The role of microorganisms in aerobic corrosion was postulated by Olsen and Szybalski (1949) as being due in part to the formation of tubercles in conjunction with microbial growth, which initiates oxygen concentration cells. This mechanism, along with others, was proposed as the cause of the worldwide problem of microbiologically associated aluminum wing tank corrosion which surfaced in the late 1950s and early 1960s. Both commercial and military aircraft were affected. Many microorganisms were reported to be present in significant numbers in the fuel tank sludge (Churchill, 1963). Recent evidence appears to indicate that organic acids, produced by fungi, were primarily involved in this corrosion (Miller, 1981).

In the 1970s a number of corrosion-induced equipment failures in the chemical process industries occurred which have now been attributed to the activities of microorganisms (Pope et al., 1984). Also, serious problems due to the activities of SRB have arisen in offshore oil operations (Hamilton and Sanders, 1986).

In addition to corrosion problems, two other problems have arisen due to bacterial growth in the legs and storage cells of offshore structures. These are the production of hydrogen sulfide, which is a serious personnel safety hazard, and the production of bacterial metabolites that give rise to accelerated concrete deterioration (Wilkinson, 1983). Biological corrosion (internal and external) of long, large-capacity, subsea pipelines, which are the present method of transmission to shore of oil and gas from offshore production fields, also appears to be a major problem (King et al., 1986a).

B. Economic Aspects

Microbial corrosion, primarily due to SRB, is a significant cause of the corrosion of underground structures, especially pipelines. Booth (1964a) estimated, for example, that at least 50% of all pipeline failures in Great Britain were due to the activities of SRB. It is also highly probable that an equal or greater percentage of failures occur in the United States, although bacterial corrosion failures are not recognized as such by many pipeline operators. It has been estimated that nearly one-half of the corrosion of steel culvert pipe in Wisconsin is due to the activities of SRB (Patenaude, 1986). In 1952, SRB were associated with the corrosion of 70% of all the seriously corroded water mains examined in the United Kingdom (Butlin et al., 1952, cited in Pankhurst, 1968).

In the petroleum industry, well casings, pumps, oil tanks, refinery equipment, and gas storage tanks may also be seriously corroded as a result of SRB activity (Iverson and Olson, 1984). An older report indicated that 77% of the corrosion occurring in one group of producing wells was due to SRB (Allred et al., 1959). The corrosion of aircraft fuel tanks has been mentioned (Miller, 1981).

Ships and marine structures have been reported to be quite vulnerable to biological corrosion. This includes ship hulls (Copenhagen, 1966; Patterson, 1951), engines (Hill, 1983), holds (Stranger-Johannessen, 1986), and fuel tanks (Klemme and Leonard,

1971) and offshore oil recovery platforms (Hamilton and Sanders, 1985; Eidsa, and Risberg, 1986) and ocean thermal energy conversion (OTEC) heat exchangers (OTEC Symposium, 1977).

In addition to the oil recovery, the gas, oil, and water supply, and the marine industries, the power generation industries (Pope et al., 1982; Bibb, 1986) and chemical and other process industries (Tatnall, 1981a; Silva et al., 1986; Wakerly, 1979) now report that microbial corrosion is of considerable significance. This corrosion has been mainly in cooling systems and heat exchangers.

Although the total actual cost of metal biocorrosion is difficult to ascertain, some recent estimates of microbial corrosion have been made. Based on a National Bureau of Standards study (1978), the national cost of corrosion of metals (replacement, prevention, and maintenance) in the United States is estimated to be 167 billion dollars in 1985. A study by the National Corrosion Service (Wakerly, 1979) in the United Kingdom indicated possible microbial corrosion involvement in 10% of cases replying to its corrosion inquires. Microbial corrosion at a similar level in the United States could account for 16–17 billion out of the 167 billion dollars. This does not include the cost of destruction, trauma, and deaths due to equipment failures such as gas pipeline ruptures.

Even in 1954, it was estimated that the corrosion of underground pipelines cost the United States 500–2000 million dollars (Greathouse and Wessel, 1954), with about 50% of these figures probably being due to microbially induced corrosion, based on the estimates given by Booth (1964a) for the United Kingdom. In 1981, the cost of corrosion just in water distribution systems in the United States was estimated to be 600 million dollars (Tuovinen et al., 1980), a considerable portion probably being of microbial origin. There are a few older reports of microbial corrosion costs in other countries. In 1967, biodamage to metals in France amounted to 1.5 billion francs (Andreyuk and Kozlova, 1981) and in 1969 the cost of microbial corrosion in Australia and New Zealand was reported to be 2.5 million and 5 million dollars, respectively (McDougall, 1969).

A table indicating corrosion rates due to cases or suspected cases of microbial corrosion in the field and laboratory has been presented by Costello (1969).

II. Corrosion

The corrosion of metals, particularly iron, is familiar to most persons as the process of "rusting." Rust is only a term referring to the corrosion product of iron or its alloys, consisting largely of hydrous ferric oxides. The corrosion of nonferrous metals is usually accompanied by the formation of their respective oxides, which vary from blue-green or red (in the case of copper) to white (in the case of zinc). Corrosion has been defined by Uhlig (1963) as the "destructive attack of a metal by chemical or electrochemical reaction with its environment." Deterioration of metals by physical means such as wear and erosion or of nonmetals such as plastic or wood is not included in the definition.

The basic cause of corrosion is the natural instability of metals in their refined forms. Metals tend, therefore, to revert to their natural states through the process of corrosion because of the free-energy change. The corrosion products may form loose films or very thin, tightly adhesive, protective films, which decrease the rate of corrosion.

A. Electrochemical Aspects

Corrosion is essentially an electrochemical process. This conclusion was first reached by Whitney (1903). The basic process of corrosion involves, therefore, a flow of electricity between certain areas of a metal surface through a solution which has the ability to conduct an electric current. More detailed reviews of the electrochemical process are to be found in books by Shrier (1976), Evans (1960), and Uhlig (1963).

In general, when a metal is immersed in water, the metal dissolves at certain sites (anodic sites), leaving behind an excess of electrons:

$$M \rightleftharpoons M^{Z+} + Ze$$

Removal of the electrons by this anodic reaction causes the reaction to be shifted to the right, thereby increasing the metal dissolution or corrosion. Two main reactions (cathodic reactions), in which the electrons are removed, involve oxygen in neutral or alkaline solutions

$$\tfrac{1}{2}O_2 + H_2O + 2e \rightarrow 2OH^-$$

and protons under acidic conditions

$$2H^+ + 2e \rightarrow 2H \rightarrow H_2$$

Areas on the metal where these electron utilization reactions occur are called cathodic sites. The products of the anode and cathode processes usually react to yield corrosion products, as in the case of iron:

$$Fe^{2+} + 2(OH)^- \to Fe(OH)_2$$

In the presence of additional oxygen, ferrous hydroxide is converted to ferric hydroxide

$$Fe(OH)_2 + \tfrac{1}{2} H_2O + \tfrac{1}{4} O_2 \to Fe(OH)_3$$

which precipitates as hydrous ferric oxide ($Fe_2O_3 \cdot H_2O$), a form of rust. Electrons also can be removed by dissimilar metal coupling or by connection to the positive terminal of a battery.

B. Formation of Corrosion Cells

Under actual conditions of use, corrosion in near-neutral environments is usually initiated through the formation of electrolytic cells. These may be established by differences in the electrical potential of dissimilar metals that are in electrical contact or differences in potential from point to point on the same metal surface. For example, if iron and copper are in contact in solution the copper ions act as electron acceptors in the dissolution (corrosion) of iron:

$$Fe \to Fe^{2+} + 2e$$
$$2e + Cu^{2+} \to Cu$$

When a metal surface is covered with a deposit, the metal outside the deposit will be accessible to oxygen while that under the deposit will be shielded from it. This results in a corrosion cell, with the formation of metal ions in the anodic region under the deposit, leading to pitting. The electrons will flow to the metal surface outside the deposit to reduce oxygen (cathodic reaction), forming hydroxyl ions. This type of cell is referred to as an oxygen concentration cell (Fig. 1) and is primarily involved in microbial corrosion. A type of corrosion similar to pitting corrosion is crevice corrosion. Crevice corrosion occurs within crevices or other shielded areas where a stagnant solution is present, whereas pitting usually takes place on a smooth metal surface, due to deposits of microorganisms on a metal surface. Other types of corrosion cells which may be due to microbial activities in deposits are various chemical concentration cells.

III. Microorganisms Involved in Corrosion

The microorganisms which have been associated with corrosion involve many genera and species. They may be divided into three groups: (1)

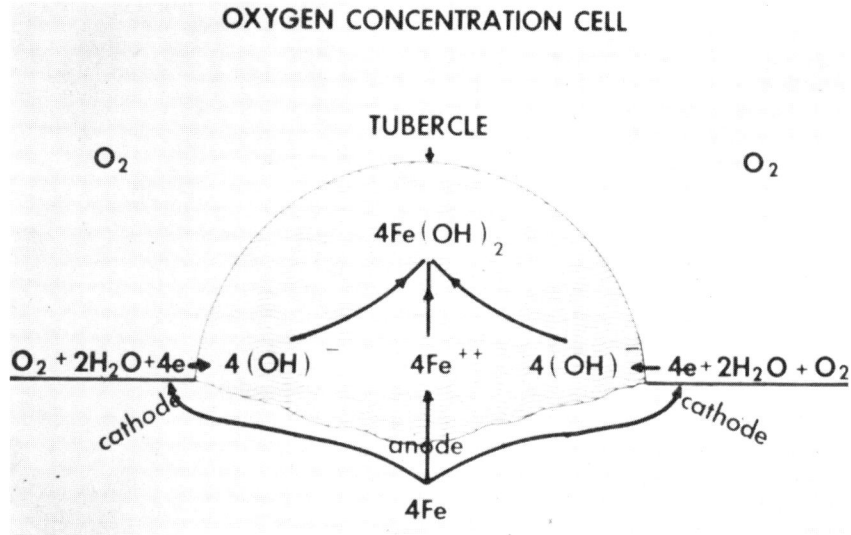

FIG. 1. Oxygen concentration cell.

bacteria, (2) fungi, and (3) algae. Many of these organisms have been firmly established as having roles in the corrosion process as a result of laboratory and field studies, whereas others have merely been isolated from suspected corrosion cases.

A. Bacteria

The most important bacteria which play a significant role in the corrosion process are those involved in the sulfur cycle (Bos and Kuenen, 1983). These include those involved in the oxidation as well as in the reduction of sulfur. Of these two groups, the SRB are the most significant bacteria found in microbial corrosion processes.

1. Sulfate-Reducing Bacteria

The SRB are a group of strict anaerobes which are taxonomically diverse but physiologically and ecologically similar. They can survive, however, for long periods of time in the presence of oxygen. They are distinguished by their ability to conduct dissimilatory sulfate reduction, using sulfate as the terminal electron acceptor and reducing it to sulfide. A few recently discovered species, such as *Desulfuromonas acetoxidans*, use sulfur instead of sulfate as the electron acceptor (Widdel and Pfennig, 1977). The biochemistry of sulfate reduction and physiology of SRB have been reviewed by Postgate (1984). Organisms in the genus *Desulfovibrio*, which contains seven species, are non-spore-forming bacteria which are spirals or curved rods (Postgate, 1984; Bergey's Manual, 1984). They have been associated with anaerobic corrosion in laboratory and field isolation studies, probably more than any other genus of SRB. Until recently, this genus and the spore-forming genus *Desulfotomaculum*, with five species, were considered to be the only two genera of SRB.

Due to the work of Widdel and Pfennig (*Bergey's Manual*, 1984) seven additional new genera of SRB have now been recognized. These include: *Desulfuromonas, Desulfomonas, Desulfobacter, Desulfobulbus, Desulfococcus, Desulfosarcina,* and *Desulfonema*. The most unusual feature of these new bacteria, aside from their morphological differences, is the wide range of carbon sources which they can utilize. Utilizable carbon sources for *Desulfovibrio* and *Desulfotomaculum* are primarily restricted to lactate, pyruvate, and malate, whereas with the new genera, CO_2 and fatty acids, from acetate to stearate, may be utilized.

2. Sulfur-Oxidizing Bacteria

In stressing the importance of anaerobic corrosion by SRB, corrosive effects of acids produced by microorganisms may be overlooked. The most significant and effective acid produced by microorganisms is sulfuric acid, which is formed by sulfur-oxidizing bacteria. Primarily these bacteria are acidophilic, aerobic chemolithoautotrophs in the genus *Thiobacillus*. The mechanism by which these motile, gram-negative organisms produce sulfuric acid from inorganic sulfur and reduced sulfur compounds has been reviewed by Purkiss (1971). *Thiobacillus thiooxidans* and a variant, *Thiobacillus concretivorus*, are the most common organisms referred to in the literature as being associated with corrosion, in addition to *Thiobacillus ferrooxidans*, which, as well as oxidizing sulfur, also oxidizes ferrous to ferric iron. This is of economic importance; because of this property, *T. ferrooxidans* is able to leach metal sulfide ores, and in the case of pyrite (FeS_2), oxidizes both the sulfur and the ferrous moiety. Bos and Kuenen (1983) have recently reviewed the sulfur-oxidizing bacteria and their relationship to corrosion and leaching.

3. "Iron" Bacteria

A third group of organisms, the aerobic "iron" bacteria, has also been associated with biological corrosion. Two types are included: the stalked bacteria in the genus *Gallionella*, and the filamentous bacteria in the genera *Sphaerotilus, Crenothrix, Leptothrix, Clonothrix,* and *Lieskeella* (Ehrlich,

1981). Both types contain chemolithotrophic autotrophs, obtaining energy from the oxidation of ferrous to ferric ions, which results in deposition of ferric hydroxide. They have been associated with the formation of tubercles, hard deposits of iron oxides, within water pipes. Filamentous iron bacteria have been associated with hollow, hemispherical tubercles in stainless-steel equipment (Tatnall, 1981b). *Gallionella* has also been reported to be associated with the corrosion of stainless steel, particularly at or near weld seams, where bacterial deposits are rich in both iron and manganese (Tatnall, 1981a,b).

4. *Miscellaneous Bacteria*

In addition to the sulfate-reducing, sulfur-oxidizing, and iron bacteria, bacteria in the genus *Pseudomonas* and pseudomonas-like organisms have been reported in connection with cases of corrosion. A strain of *Pseudomonas* isolated from corroded pipe systems carrying crude oil was found to reduce ferric iron to soluble ferrous iron, thus continually exposing a fresh surface to a corrosive environment (Obueke et al., 1981a). *Pseudomonas* species are most prevalent in industrial water environments along with several other slime-forming bacteria, where their primary role appears to be in colonizing metal surfaces, thereby creating oxygen-free environments which harbor SRB. All types of organisms which have this capacity may be considered potentially corrosive organisms.

A variety of hydrogenase-positive, photosynthetic, and nonphotosynthetic bacteria have been tested in the laboratory for their corrosive effect by the cathodic polarization technique (Mara and Williams, 1971, 1972a) and by measurement of iron loss (Umbreit, 1976) following formation of rust. Only small effects were noted. In laboratory studies, however, several heterotrophic bacteria which form H_2 and CO_2 (Frenzel, 1965, 1966) and acids (Ehlert, 1967) have been reported to play an important role during the corrosion of iron. Moreau and Brisou (1972) have classified heterotrophic aerobic and facultative bacteria into four groups, based on their proteolytic,nitrate-reducing, and carbohydrate-breakdown properties, with respect to the corrosion of copper and nickel. Some groups inhibited corrosion and others caused limited or extensive corrosion.

B. FUNGI

From a biological corrosion viewpoint, one of the most significant fungi is *Cladosporium resinae*, which is involved in the corrosion of aluminum integral to fuel tanks of subsonic aircraft, leading to wing perforation and loss of fuel. This organism grows at the water/fuel interface at the bottom of fuel tanks and utilizes components of the fuel (C_3–C_{16} alkanes) and inorganic constituents, dissolved in the water, for nutrients (Miller, 1981; Tiller, 1982). It is now believed that the aluminum corrosion is caused by carboxylic acid production by C. resinae (Miller, 1981). In supersonic aircraft fuel tanks, an albino mutant strain of Aspergillus fumigatus has been shown to be the primary corrosive agent due to its ability to grow at the higher temperatures in the tanks (Hill and Thomas, 1975).

In an earlier report, a Cerostomella sp. was reported to have caused the corrosion of steel and aluminum-clad aircraft parts that were still inside their original wooden packing cases (Copenhagen, 1950). Later, Al'bitskaya and Shaposhnikova (1960) demonstrated the corrosion of copper, aluminum, and steel using five acid-producing species of fungi: Aspergillus niger, Aspergillus amstelodami, Penicillium cyclopium, Penicillium brevicompactum, and Paecilomyces varioti. A recent report describes the corrosion of the interior of a ship's hold by fungi (Stranger-Johannessen, 1986).

C. Algae

Algae are important fouling organisms and although there are relatively few reports of direct corrosion by algae, they would appear to have the potential for inducing corrosion by virtue of their role in production of oxygen, corrosive organic acids (Prescott, 1968), and nutrients for other corrosive microoganisms, as well as their role in slime formation (concentration cells). Das and Mishra (1986) have reported on the corrosion of various types of welded mild steel and 304 stainless-steel samples by two species of blue-green algae (Nostoc parmelioides and Anabaena sphaerica) and a species of red algae (Graciollasia sp.) from the Bay of Bengal. Cathodic polarization studies on three strains of hydrogenase-positive Chlorophyta and three strains of hydrogenase-positive Cyanophyta indicated that the organisms were able to utilize cathodic hydrogen (Mara and Williams, 1972a). It has, however, been found that under healthy microalgal mats composed of Oscillatoria sp., colonial diatoms, and Enteromopha sp., the pH is raised to high values, which tends to reduce the corrosion rate (Edyvean and Terry, 1983). In areas where there is decay of the mat the pH was lowered, probably due to the production of corrosive organic acids, causing differential pH corrosion cells as well. Conditions were also found to be rendered favorable for the growth of SRB.

IV. Mechanisms of Microbial Corrosion

The mechanisms by which microorganisms induce corrosion essentially involve the basic electrochemical mechanisms of corrosion previously discussed, namely, the removal of electrons via oxygen or hydrogen ions. The one exception is the mechanism for the corrosion of iron in the absence of oxygen and at neutral or near-neutral pH values. A number of theories have been proposed to elucidate this phenomenon, the foremost one being the classical cathodic depolarization theory.

A. Cathodic Depolarization

1. Classical Theory

Based on the facts that anaerobic corrosion in deoxygenated soils was associated with hydrogen sulfide, that hydrogen sulfide was produced by SRB, and that these bacteria could utilize molecular hydrogen for the reduction of sulfate sulfur (Stephenson and Stickland, 1931), von Wolzogen Kühr and van der Vlugt (1934) proposed a cathodic depolarization theory to account for this corrosion. The essential step in this theory involves the removal of hydrogen (cathodic depolarization) by the hydrogenase system of the SRB (Fig. 2). Evidence for and against this theory has filled much of the literature on microbial corrosion. Presently, the bulk of evidence does not appear to support it.

CATHODIC DEPOLARIZATION THEORY
of von Wolzogen Kühr and van Der Vlugt

I $8H_2O \longrightarrow 8OH^- + 8H^+$

II $4Fe \longrightarrow 4Fe^{++} + 8e$ (anode)

III $8H^+ + 8e \longrightarrow 8H$ (cathode)

IV $SO_4^{--} + 8H \xrightarrow{\text{(bacteria)}} S^{--} + 4H_2O$ (cathodic depolarization)

V $Fe^{++} + S^{--} \longrightarrow FeS$ (anode)

VI $3Fe^{++} + 6(OH)^- \longrightarrow 3Fe(OH)_2$ (anode)

$4Fe + SO_4^{--} + 4H_2O \longrightarrow FeS + 3Fe(OH)_2 + 2(OH)^-$

FIG. 2. Cathodic depolarization theory of von Wolzogen Kühr and van der Vlugt (1934).

According to the theory, the ratio of corroded iron to iron sulfide should be 4:1, whereas it actually is found to vary from 0.9:1 to almost 50:1 (Tiller, 1982). In addition, the corrosion products of iron oxide or hydroxide are seldom found. Initially, studies by Booth and Wormwell (1962), using batch cultures of several hydrogenase-positive strains of SRB, indicated a direct relation between hydrogenase activity and corrosion rate. Later, when semicontinuous cultures (Booth et al., 1965) were used, evidence in conflict with these results was obtained. A film of sulfide was found to form over the iron surface, giving some protection against corrosion, when the organisms were growing in a lactate–mineral salts medium containing sulfate. This film was found later to break down and the resulting corrosion rates showed no simple relationship to the hydrogenase activity of the organism, the greatest corrosion rates being obtained with a halophilic organism of with low hydrogenase activity.

Initial results by Iverson (1966) seemed to provide support for the depolarization theory. He demonstrated the dissolution of iron, on an agar surface (containing benyl viologen; BV), from a mild steel electrode (anode) electrically connected to a similar electrode (cathode) in contact with a culture of hydrogen-grown, hydrogenase-positive strain of *Desulfovibrio*. The *Desulfovibrio* oxidized cathodic hydrogen and transferred the electrons to the redox dye, BV (instead of sulfate), which served as an electron acceptor. A small cathodic depolarization current density (1 $\mu A/cm^2$) was recorded which was not nearly enough to account for the high rates of corrosion observed in the field. Similar results were obtained by Booth and Tiller (1968). They used a potentiostatic technique (to measure depolarization current from the cathode, maintained at various fixed potentials) and washed, hydrogen-grown, resting cells in an inert buffer, with BV again as the electron acceptor.

Costello (1974), however, has criticized the use of BV as an electron acceptor, stating that hydrogenase-positive organisms not only catalyze the uptake of hydrogen but also the liberation of hydrogen by reduced BV. He based this statement on a paper by Peck and Gest (1956) that described a procedure for the assay of bacterial hydrogenase. In this procedure, however, methyl viologen was used instead of BV, as "virtually no evolution is observed with hydrosulfite plus reduced BV."

Further results, however, by Iverson (1968) indicated that although redox dyes could be used as electron acceptors in his system, sulfate did not act as an electron acceptor when it replaced BV (i.e., no ferrous iron was produced at the anode and no cathodic depolarization current could be observed). These results were confirmed by Costello (1974) as evidenced by his polarization curves (current versus potential) in the negative or cathodic direction. Hardy (1983), using the respiration of [^{35}S] sulfate as an indicator to monitor cathodic hydrogen utilization, observed only a transient removal of hydrogen and concluded that removal of cathodic hydrogen is unlikely to be the dominant corrosion mechanism. Hardy suggested that the transient hydrogen utilization may have resulted from the poisoning of the atomic-to-molecular hydrogen reaction by the traces of sulfide, molecular hydrogen being the substrate for the hydrogenase system rather than atomic hydrogen, as originally proposed in the theory.

The effect of pressure on the corrosion of ingot iron by marine isolates of SRB was investigated by Willingham and Quinby (1971) and of mild steel by Herbert and Stott (1983). Willingham and Quinby observed that pressures up to 200 bar increased the corrosion of mild steel; higher pressures were inhibitory. Herbert and Stott, however, using SRB isolates from the North Sea, observed high corrosion rates (weight loss) of mild steel at 400 bar (30°C). Because at this pressure no hydrogenase was present and sulfide concentrations were relatively low, a new mechanism appeared to be operating.

2. Alternative Mechanisms

In addition to the classical mechanism, the role of FeS as a depolarizer was proposed by Booth et al. (1968). Their cathodic polarization curves of cultures of SRB growing without sulfate (substituted with fumarate), with and without added ferrous sulfide (FeS), indicated polarization of the hydrogen by both the bacterial hydrogenase system and the solid ferrous sulfide. The FeS may be formed as a film on the surface of the iron in actively growing cultures of SRB low in soluble iron (Booth et al., 1967c), or as bulk FeS in cultures high in soluble iron (King et al., 1973a). In the former case, the film usually inhibits corrosion, but it may break down, usually with an increase in the corrosion rate (Booth et al., 1965; Iverson and Olson, 1983). The primary films appear to be composed of mackinawite (FeS_{1-x}) and siderite ($FeCO_3$), the latter being protective (Tiller, 1983b). The interrelationships between the iron sulfides are indicated in a paper by Tiller (1983b). In the case where soluble iron has been added to the culture, forming bulk iron and preventing film formation, the corrosion rates are very high (King et al., 1973b). King concluded from this study that FeS was the major corrosive agent in the semicontinuous culture system. From a literature review, he also showed that the logarithm of the corrosion rate was directly related to the logarithm of the soluble iron (micromolar) in the media.

In other studies by King et al. (1973a) the stimulation of corrosion by chemically prepared iron sulfides was found to decrease with time, but could be restored in the presence of SRB (Tiller, 1982). Earlier, King and Miller (1971) postulated that the role of bacteria could be to depolarize (remove molecular hydrogen) the iron sulfide in contact with the steel or to bring fresh iron sulfide constantly into contact with the steel by cellular movement (Miller, 1981).

Costello (1974) found that if cell-free centrifugates of SRB were treated to scrubbing by nitrogen, the resulting cathodic polarization curves showed no depolarizing activity, as contrasted to the centrifugates before scrubbing. He concluded that the depolarizing agent was a gaseous species, which he assumed to be hydrogen sulfide, because the addition of hydrogen sulfide back into the degassed cultures showed the same types of polarization curves as before degassing.

B. Concentration Cell Formation

A wide variety of organisms, both microorganisms and macroorganisms, are able to colonize immersed metals. At the present time, investigation of this process with respect to the attachment and metabolic activity of microorganisms is a very active field. For reviews in this area, the papers by Characklis and Cooksey (1983) and Costerton et al. (1981) are excellent. The attachment to a metal surface (biofilm formation) by aerobic organisms depletes the oxygen concentration at the metal surface, thereby resulting in an oxygen differential and establishing corrosive, oxygen-differential cells. The interior surfaces of flow channels and pipes in heat exchangers, for example, may become blocked by the accumulated mass of macro- and microorganisms, resulting in a problem referred to as biofouling. The oxygen concentration under the biofilms may be so depleted that SRB may be established (Hamilton, 1985). Estimates of the depth of biofilms of bacteria and their associated extracellular polymers, which produce anaerobiosis under the films, vary from 10 (Costerton and Gessey, 1979) to 100 μm (Hamilton, 1985).

Almost any type of microorganism which can colonize a surface may therefore be considered as a potential corrosion initiator. These microorganisms would include filamentous blue-green and green algae and bacteria such as *Pseudomonas* spp., *Aerobacter* spp., *Flavobacterium* spp., *Gallionella* spp., and *Sphaerotilus* spp., in addition to fungi such as *Aspergillus* spp. and *Cladosporium* spp. Techniques are being developed to identify the microorganisms within a biofilm *in situ*. These include the identification of lipids characteristic (unique fatty acid profiles) of fungi, algae, and bacteria, including the SRB (White, 1983; Taylor and Parkes, 1983), and fluorescent antibody techniques (Pope, 1982). Techniques have also been developed to measure the reactivity of microorganisms (White et al., 1986), particularly sulfate reducers, in situ (Hamilton, 1985).

In addition to the blockage of water in pipelines by biofouling, restriction of potable water flow in water-distribution pipelines may result in the formation of tubercles. The effects of fouling on corrosion as well as the effects of corrosion on microorganisms have been discussed by Gerchakov and Udey (1982). Tubercles are deposits of magnetite (Fe_3O_4) and goethite [α-FeO(OH)] in water pipes (Ainsworth et al., 1978; Hutchinson and Ridgway, 1977; Olsen and Szybalski, 1949) which shield the surface of pipes from oxygen, thus creating oxygen concentration cells (Fig. 3). The small anodic area beneath the tubercle often corrodes severely, resulting in perforation of the pipe.

C. Corrosive Metabolic Products

1. Volatile Phosphorus Compound

In the experiment previously described, sulfate was not found to replace BV as an electron acceptor for the cathodic depolarization of iron (Iverson, 1968). A black deposit of an iron compound was found under the electrode in contact with the Desulfovibrio cells (normally the cathode when BV was used as the electron acceptor). Further investigation revealed the black corrosion product to be amorphous iron phosphide (Fe_2P) after heating in a vacuum oven (Iverson, 1968).

Previous reports by many investigators (King et al., 1973b) stated that the rate of anaerobic corrosion was related to the ferrous ion concentration in the medium; the greater the iron concentration the greater the corrosion rate. Extensive corrosion of steel (1250 mg/dm^{-2} day^{-1}) was obtained under anaerobic conditions in a filtrate of a marine culture of Desulfovibrio from which all the free sulfide ions were removed by addition of ferrous ions and the resulting iron sulfide and the bacterial cells were removed by Seitz filtration (Iverson, 1974, 1981). These results indicated that neither iron sulfide, sulfide ions, nor bacterial cells were necessary for the corrosion. Furthermore, the main corrosion product was again found to be amorphous iron phosphide. Further results indicated that the corrosive agent was a volatile phosphorus compound (Iverson and Olson, 1983) produced by the SRB. A similarly acting, corrosive, volatile, phosphorus compound was also produced by the action of hydrogen sulfide on certain crystals of disodium phosphate

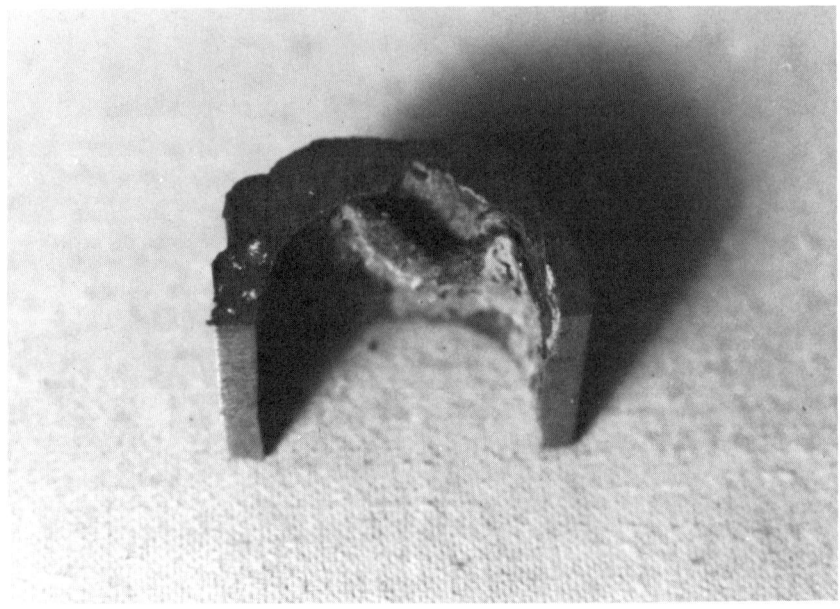

FIG. 3. Corrosion pit in section of iron (mild steel) pipe underneath tubercle.

and sodium hypophosphite (Iverson et al., 1985). Iverson et al. (1986) proposed that the stimulation of corrosion by ferrous ions as noted by King et al. (1973b) was due to the prevention of iron sulfide film formation by the precipitation of sulfide ions as bulk iron sulfide in solution, thereby permitting the corrosive phosphorus compound to come into contact with the bare iron surface and initiate corrosion. Also, the breakdown of a preformed iron sulfide film would permit the phosphorus compound access to the iron surface and initiate corrosion. Initiation of corrosion after film breakdown was reported by King et al. (1973b), Booth et al. (1965), and Iverson and Olson (1983). The observation by Herbert and Stott (1983) that high anaerobic corrosion rates at high pressures (400 bar), where hydrogenase activity was absent and sulfide concentrations were low, may have been due to this phosphorus compound.

2. *Acids*

Both inorganic and organic acids produced by microorganisms may be highly corrosive to metals. The main inorganic acid involved in metal corrosion is sulfuric acid produced by acidophilic, colorless, sulfur-oxidizing bacteria. These bacteria may be found in environments where reduced sulfur compounds are present, and if oxygen is available, very low pH values (pH 2–5) may result. The primary organisms involved are *Thiobacilus thiooxidans* and *Thiobacillus ferrooxidans*, the latter organism being more frequently associated with acid mine drainage. Several oxidation reactions of sulfur and sulfur compounds have been outlined by Cragnolino and Tuovinen (1984). Baru et al. (1982) obtained results which indicated that *T. thiooxidans* produced active corrosive metabolic products, in addition to sulfuric acid, which intensified the corrosion of steel.

Failures of iron pipes due to this mechanism have been reported (Booth, 1971). A case of sulfuric acid corrosion of copper pipe in bog soil used as fill soil has also been reported (Schwerdtfeger, 1967). Since oxygen is required in the bacterial oxidation of sulfur and its compounds, cases of this corrosion in the soil are not nearly as common as anaerobic corrosion. The degradation of concrete sewers where there is an abundance of hydrogen sulfide is probably the most important problem caused by the sulfur-oxidizing bacteria (Tiller, 1983b). Deterioration of building stone and structures such as cooling towers and monuments by these organisms is also a serious problem (Tiller, 1983b). Fatty acids produced by marine bacteria may also contribute to the deterioration of concrete used in storage cells and platform legs of offshore structures (Morgan et al., 1983).

Corrosion by organic acids is primarily due to the activities of fungi, although some cases of corrosion by bacterially produced organic acids have been reported. In the corrosion of aircraft aluminum alloys (2024 and 7075) by *Cladosporium resinae*, carboxylic acid production by this organism appears to be one responsible agent. Citric, isocitric, cis-aconitic, and α-oxoglutaric acids were shown to accumulate during growth in a Bushnell–Haas medium with kerosene as a carbon source (McKenzie et al., 1977). Aluminum alloy 2024-T3 was found to corrode most rapidly in α-oxoglutaric acid, whereas the alloy 7075-T6 corroded most rapidly in citric acid, with corrosion rates being similar to those observed in the fungus cultures. The results of Salvarezza et al. (1983) also showed that the metabolites of *C. resinae* caused a shift in the pitting potential of aluminum alloys.

In addition to the corrosion of aluminum, steel and copper (wire) have been reported to be corroded by organic acids of fungal origin (Al'bitskaya and Shaposhnikova, 1960). The corrosion effects of organic acids, including glycolic, tricarboxylic, amino, and fatty acids, on several metals were reported by Staffeldt and Calderon (1967) and Burns et al. (1967).

Organic acids from a number of gram-negative bacteria were found to cause extensive corrosion of iron (Ehlert, 1967). Corrosion in a sugar beet factory was reported to be due to acids from *Lactobacillus delbrueckii* (Allen et al., 1948), and in soil, corrosion of jute wrapped-metal was due to acetic and butyric acids from cellulose-decomposing bacteria (Coles and Davies, 1956).

3. Sulfur and Sulfur Compounds

At or near neutral pH values, hydrogen sulfide produced by SRB, mainly in the form of HS⁻ ions (Ray, 1982), reacts with iron to form a partially protective film of iron sulfide that may afford varying degrees of corrosion protection, as mentioned previously. Under anaerobic conditions and at low pH values, were the hydrogen sulfide may be unbuffered or where the acidity may be due to organic acids such as naphthenic acids from petroleum, hydrogen sulfide will corrode iron with the release of hydrogen. A wide range of equipment in the petroleum industry is attacked by this mechanism (Iverson and Olson, 1984).

In addition to corrosion, other types of damage to steel may result. Hydrogen (atomic), formed by the action of hydrogen sulfide on iron, may be responsible for embrittlement of hardened steel parts as well as blistering. Hydrogen embrittlement may occur when atomic hydrogen diffuses into steel that is under stress, altering the metal characteristics and causing crack initiation. Hydrogen blistering is believed to occur when the atomic hydrogen, reacting with itself, forms hydrogen gas, which results in tremendous pressures within the steel structure and results in blistering (Tuttle and Kane, 1981). A variety of sulfur species are also able to induce intergranular stress corrosion cracking (IGSCC) of sensitized austenitic stainless steels at ambient temperature (Cragnolino and MacDonald, 1982).

Hydrogen sulfide may also be spontaneously or biologically oxidized to elemental sulfur in the presence of oxygen, or anaerobically by biological photosynthetic oxidation, e.g., green and purple sulfur bacteria (Bos and Kuenen, 1983). Elemental sulfur has been reported to be corrosive to mild steel under aerobic as well as anaerobic conditions (Farrer and Wormwell, 1953; MacDonald et al., 1978). In a more recent paper, Schaschl (1980) considers the corrosion by sulfur in deaerated neutral solutions to be more important than the action of SRB. Maldonado and Boden (1981) determined that elemental sulfur formed H_2SO_4 and H_2S in demineralized water. Mercaptans produced by a variety of organisms are also corrosive to certain metals. Iron sulfide previously was suggested as a corrosive agent for iron. Tiller and Booth (1968) suggested that ferrous sulfide may also be involved in the corrosion of aluminum.

4. Ammonia and Amines

The corrosive action of ammonia on copper and brass is well documented (Shrier, 1976). Ammonia is also a very ubiquitous product of microbial metabolism. Probably, the first report of attack on copper alloys by ammonia produced by microorganisms was by Bengough and May (1924). Primary, secondary, and tertiary amines have also been demonstrated to attack copper and nickel (Guillaume et al., 1970).

5. Hydrogen

Hydrogen atoms may accumulate on the surface of some ferrous alloys because of poisonous compounds (such as hydrogen sulfide) that retard the rate of formation of molecular hydrogen. This would be considered good evidence against the cathodic depolarization theory, as the SRB are believed to utilize only molecular hydrogen. The hydrogen atoms may enter the alloy and cause hydrogen embrittlement (loss of ductility) and hydrogen cracking (spontaneous cracking). In this connection, Walch and Mitchell (1986) have indicated that small amounts of hydrogen may be absorbed by metals beneath films of hydrogen-producing bacteria such as *Ruminococcus albus*.

D. PROTECTIVE FILM DISRUPTION

Protective films may be natural or artificial. Metals in natural environments form films of corrosion products (usually oxides, in the presence of air) that may afford corrosion protection in varying degrees. The oxide film on aluminum, for example, provides considerable protection but the oxide film on iron is usually poorly protective. Artifical films may be various man-made paints or organic coatings of various types. Both kinds of films have been reported to have been broken down by biological action.

Steel H-piles in the ocean, for example, were reported to have been destroyed by the action of purple sea urchins, which removed the partially protective oxide films through their abrasive action (Muraoka, 1968). Recently, a report appeared on the destruction of protective ferric coatings by a ferric-iron-reducing pseudomonad that was thought to be responsible for high levels of corrosion (Obuekwe et al., 1981b). The organism was thought to reduce the insoluble ferric film to a soluble ferrous iron, thus causing the protective film to be destroyed. It has already been mentioned that the breakdown of iron sulfide films in cultures of SRB is usually followed by high corrosion rates. Acidic metabolites produced by *C. resinae* were found by Salvarezza et al.

(1983) to facilitate the breakdown of the passive aluminum oxide film by chloride ions, thus decreasing the pitting potential. Bacterial slimes on a metal surface were, on the other hand, suggested to assist in preserving the oxide films by cementing them to the surface (Gerchakov and Udey, 1982).

Various protective organic coatings applied to buried pipes (Harris, 1960) and aluminum fuel tanks (Miller et al., 1964) have been broken down due to microbial utilization of the coating components as sources of nutrients.

E. BREAKDOWN OF CORROSION INHIBITORS

Various corrosion inhibitors are quite commonly employed in closed aqueous systems. Many of them can be utilized or transformed by the microorganisms present. Some organic corrosion inhibitors such as diamines and aliphatics can be utilized as nutrients, for example. Nitrates, aluminum corrosion inhibitors, can be reduced to nitrites (Blanchard and Goucher, 1967). This microbial transformation was found to lead to a decrease in the pitting potential of aluminum (Salvarezza et al., 1983).

F. COMBINATION OF MECHANISMS

Under natural conditions, a number of corrosion mechanisms probably operate, simultaneously or in succession. In the case of biofilms, oxygen concentration cells may operate along with proton gradients produced as a result of the metabolic activities of the microorganisms in the biofilm. As regions under the film become depleted in oxygen, SRB may become established, which, in addition to accentuating the oxygen concentration cell effect, may further stimulate corrosion by elaboration of the corrosive phosphorus compound.

In environments where there is considerable sulfate-reducing activity, the sulfides may be oxidized to corrosive, elemental sulfur or sulfuric acid, if the environments become oxygenated.

Other examples of proposed multiple mechanisms include the laboratory study by Little et al. (1986) on the corrosion of nickel 201 by colonization with a thermophilic bacterium. They suggested that the corrosion was due to three simultaneous activities: (1) the creation of differential aeration cells, (2) the production of acidic metabolites (isobutyric and isovaleric acids), and (3) the chelation of heavy metals, which results in galvanic currents. As a result of laboratory studies, Videla (1986) proposed three mechanisms for the corrosion of aluminum and aluminum alloys by C. resinae: (1) a local increase in the proton concentration from organic acid production, (2) the metabolic production of substances that decrease the stability of the passive film on the metal surface, and (3) a diminution of the concentration of corrosion inhibitors in the media by microbial utilization.

V. Microbial Corrosion of Ferrous Alloys

The microbial corrosion of ferrous alloys, such as mild steel, in oxygenated seawater has been difficult to establish over short periods of time. Smith and Compton (1973) exposed 1016 mild steel (carbon steel) specimens to flowing, nonrecirculated, nonsterile seawater. The sterile seawater was passed through a membrane filter and a small portion was recirculated. Corrosion rates were measured by a polarization technique (polarization resistance measurement). No significant difference was found in the corrosion rates of the specimens exposed to sterile and nonsterile water (7.0 and 7.4 mils per year). Although slime-forming aerobic marine microorganisms were found on the oxide films, they did not form coherent slime accumulations on the steel surface.

In long-term exposure (up to 16 years) of structural mild steel (1020) panels, Southwell and Bultman (1972) found that the corrosion rate, initially high (10 mils per year) during the first year, dropped to a low value of less than 0.5 mil per year due to an accumulation of fouling. After 4 years, anaerobic corrosion by SRB became established and a steady state of 2.7 mils per year was reached between 4 and 16 years of exposure. At the end of 16 years the deepest penetration was found to be only three times the average penetration.

These ferrous alloys have received the most attention, however, with respect to microbiological corrosion, due to their universal use in underground structures. In the case of cast iron, the iron is dissolved away leaving a soft residue of graphite, giving the appearance of an intact pipe. At times the corrosion rates may be extremely high. Booth (1964b) reported rates as high as 250 mils per year. In a study of the corrosion of cast iron and mild steels containing increasing amounts of carbon, Mara and Williams (1972b) found that in a minimal-iron nutrient medium a linear relationship existed between the weight loss of the steel specimens and their carbon contents, the cast iron corroding faster than the steel specimens. In an iron-rich medium, however, the cast iron and steel specimens were found to corrode at the same rate.

A study of the corrosion of grey cast iron and a new spun ductile iron by SRB in aqueous and soil media was reported by King et al. (1986b). Spun ductile iron pipes were used by the United Kingdom water industry in the 1960s instead of spun grey iron pipe (common type of cast iron). The corrosion behavior of the two types of cast iron was found to be about equal, but their properties after corrosion were different (*Science News*, 1985). The ductile iron, containing a spheroidal structural graphite, was found to crumble whereas the grey iron, containing a flaked graphite matrix, was found to maintain its integrity.

Suspected cases of microbial corrosion of stainless steel have been reported (Tiller, 1983a; Thorpe, 1976; Kobrin, 1976; Tatnall, 1981a,b). Stainless steels are alloys of iron containing 12–30% chromium to provide corrosion resistance. Although they are not usually resistant to reducing environments they are quite corrosion resistant under oxidizing conditions. Areas depleted of oxidants such as crevices or under slime deposits are therefore quite susceptible to corrosion. In the chemical-processing and paper industries the most commonly used stainless steel is of the austenitic type (300 series), which contains 8–35% nickel and 12–25% chromium. Very few studies on the biological corrosion mechanisms of these alloys have been carried out, so the underlying processes remain conjectural.

In a number of cases, metal-oxidizing bacteria such as *Gallionella* spp. have been found in association with corrosion pits in stainless steels (Tiller, 1983a; Kobrin, 1976; Tatnall, 1981a). Although these organisms (*Gallionella*) are not associated with the oxidation of manganese, their deposits in the form of cones are rich in iron, manganese, and chlorides (Tatnall, 1981b). In many cases corrosion of the type produced by these bacteria has been associated with weld seams (Kobrin, 1976). Rust-colored streaks from *Gallionella* deposits on sidewall seams have been noted (Tatnall, 1981b). It has been postulated that the accumulation of chlorides, arising from a number of mechanisms which are known to be highly damaging to the normally protective film, could instigate the corrosion (Pope et al., 1984).

SRB have also been associated with the corrosion of stainless steels (Tiller, 1983a; Kobrin, 1976, Tatnall, 1981a), usually in association with *Gallionella* spp. or organisms such as *Pseudomonas* and *Sphaerotilus* spp. It is thought that slime-forming bacteria, including aerobic bacteria and the iron bacteria, deplete the oxygen and create conditions for classical crevice corrosion and that the SRB corrode the stainless steel by one or more of the postulated mechanisms (Tatnall, 1981b).

VI. Microbial Corrosion of Nonferrous Metals and Alloys

A. COPPER

It has been believed that copper ions are quite toxic to any organisms that might cause corrosion of copper, hence the microbiological corrosion of copper has been regarded as insignificant (Leidheiser, 1971). In actuality, this is not always true, as copper resistant organisms have been found in association with corrosion of copper (Rogers, 1948). Ammonia, produced by many microorganisms, has long been known to be corrosive to copper and brass. Probably the first report suggesting the relation between bacterially produced ammonia and the pitting of cooper condenser tubes in power stations was by Grant et al. in 1921. Bengough and May (1924) also reported on the corrosion of copper condenser tubes by microbially produced ammonia.

The catastrophic effect of an acid fill (originally anaerobic) soil on copper pipe has been mentioned (Schwerdtfeger, 1967), the sulfuric acid being produced by the action of *Thiobacillus* spp. on sulfides. SRB also have been reported to have a corrosive effect on underground pipes of copper and its alloys (Gilbert, 1946; Alanis et al., 1986). Rogers (1948), in a study of the corrosion of copper alloys in seawater, found that corrosion of 70–30 Cu–Zn brass was much more severe when nutrients were added. Corrosion was in the form of deep pits under colonies of copper-tolerant bacteria. A large percentage of the seawater samples was found to contain high levels of organic sulfur compounds, thought to be reduced to mercaptans and then oxidized to sulfides. It was found, however, that addition of bacteria to sterile seawater systems containing a mercaptan and disulfides (which accelerated corrosion by themselves) further increased the rate of corrosion.

Several reports have indicate that polluted or "putrid" seawater, rich in sulfides, accelerates the corrosion of copper-based alloys used in heat exchanger tubing and ship seawater piping (Rowlands, 1965; Eiselstein et al., 1983; Efird and Lee, 1979; Syrett et al., 1979; Schiffrin and Sanchez, 1985). The corrosion of copper–nickel alloys when exposed to putrid seawater was found to be accelerated (Efird and Lee, 1979; Eiselstein et al., 1983). The corrosion in deaerated, sulfide-polluted seawater was believed to arise from a shift in the corrosion potential to active values, in which hydrogen evolution becomes the main cathodic reaction (MacDonald et al., 1979). In the presence of oxygen, the sulfur of the sulfide film is believed to be oxidized to its oxidation products (viz., sulfur, sulfate, sulfite, and polysulfides), thereby increasing the corrosion rate (Syrett et al., 1979).

In bacterial studies of copper corrosion, Engel (1969) found that about half of the bacteria (11 types) isolated from aircraft fuel tanks, fuel storage tanks, and air and water samples caused slight corrosion of copper alloys. No viable organisms were detected, however, after 6 months. Guillaume et al. (1973a) found, in contrast, that the corrosion of copper in fresh water and seawater inoculated with several types of bacteria was inhibited and that corrosion increased after the death of the bacteria. Utilizing copper-resistant bacteria (*Pseudomonas* sp., *Micrococcus* sp., and *Corynebacterium* sp.) isolated from gelatinous deposits on the inside walls of copper alloy condenser tubes in contact with seawater, Schiffrin and Sanchez (1985) found that only the possibly copper-tolerant, film-forming *Pseudomonas* sp. was capable of enhancing the corrosion rate of 90–10 Cu–Ni and Al–bronze alloys (about 20-fold over that in the sterile seawater medium).

B. ALUMINUM

Aluminum is a very active metal which would corrode very rapidly except for a protective oxide (passive) film. If the film breaks down, dissolution of the aluminum is rapid. The association of microorganisms with the corrosion of aluminum began in the 1960s with the use of "wet-wing" aircraft (kerosene-type fuels were stored in the wings, i.e., integral fuel tanks) on which piping and pumps were used to maintain trim during flight by fuel transfer. Water is usually present in integral fuel tanks even though water drainage points are provided. It may arise from (1) condensation of moisture in the air, (2) condensation of water miscible in the fuel, and (3) fuel-handling techniques (immiscible water). The organisms involved in corrosion may grow in the form of mats or sludge on the bottom of the tanks. Corrosion may be associated with the development of volcano-shaped tubercles with frequent evolution of gas.

A number of microorganisms have been isolated from the sludges and tubercles, including bacteria and fungi (Churchill, 1963). The most commonly identified species have been *Cladosporium* spp., *Pseudomonas* spp., and *Desulfovibrio* spp. (Churchill, 1963; Miller et al., 1964). Although Iverson (1967) isolated SRB from aluminum corrosion pits and proposed an oxygen concentration cell mechanism to explain the corrosion, it was Tiller and Booth (1968) who demonstrated the corrosion (pitting) of aluminum by three strains of SRB. The highest weight loss rate they observed (due to *Desulfovibrio salexigens*) was 100 times the rate of the sterile controls. Under a pressure of 200 bar only slight aluminum corrosion was noted, however, in some instances (Willingham and Quinby, 1971). Oxygen concentration cells, formed beneath fungal mats, were believed by Miller et al. (1964) to have a sufficient potential difference (up to 60 μV) to cause rapid pitting of aluminum alloys.

only those areas which the coating does not cover, such as small pinholes.

D. INHIBITORS

In closed systems such as tanks and recirculating cooling-water systems, the use of chemicals both to inhibit corrosion as well as to inhibit microbial growth may be employed. Discussions of anticorrosive chemicals may be found in the many books on general corrosion.

Presently, there are a number of biocides or biostats in use. These include oxidizing agents such as chlorine and ozone, phenolics, aldehydes, metal organic compounds such as organotins, heavy metal salts, and quaternary ammonium compounds.

Saleh et al. (1964) have listed the effect of some 200 biocides and bacteriostats on SRB. A list of potential compounds with microbiocidal/microbiostatic activity as well as other measures that may be taken to control biological corrosion in oil field equipment has been outlined (National Association of Corrosion Engineers, 1972). A recent assessment of some of these compounds for use in the oil recovery system is given by Bessems (1983). A review of commercial biocides for use in industrial water systems that are subject to microbial corrosion has been published (McCoy, 1974). In aircraft fuel tanks, organoborane compounds, potassium dichromate, and glycol monomethyl ether have been reported to be effective in preventing fungal corrosion (Miller, 1981). The latter compound is also an icing inhibitor which, when added to the fuel, partitions to the water phase to form inhibitory concentrations. A review of biocides for use in cutting oil and other types of oils has been published by Hill (1984).

In general, the control of microorganisms that form films that result in potential corrosion is much more difficult and requires much higher concentrations of inhibitory agents than the control of microorganisms in the aqueous phase (Costerton, 1983). Use of several compatible inhibitory compounds in rotation or simultaneously may be more successful than continued use of a single compound, since the likelihood of microbial tolerance may be decreased. The compounds should also be compatible with the corrosion inhibitor, if used. A reported case of successful corrosion control in a cooling-water system was obtained by employing an organic film corrosion inhibitor, a polyacrylate/phosphonate dispersant, and a combination of two biocides used simultaneously (Honneysett et al., 1985).

VIII. Conclusions

It is evident, from the abundance of literature on microbially induced corrosion and from various assessments of its economic effect, that microorganisms probably play a significant role in the corrosion process. This conclusion is further buttressed by good evidence that corrosion is inhibited by the use of microbial inhibitory compounds (viz., aluminum fuel tank corrosion).

Much of the evidence is suggestive, however. Relatively few reports present direct evidence, that is, evidence in accordance with Koch's postulates, namely (1) isolating the organisms from a corroded area in the field, (2) reproducing the corrosion under identical conditions in the laboratory using appropriate sterile controls, and (3) reisolating the identical organisms from the corrosion site in the laboratory. The most difficult postulate is the reproduction of corrosion under identical conditions as first noticed in the field. More often than not, the field conditions are not known and the corroded sample, usually well dehydrated, is all the evidence available. In other instances, where field conditions do exist, it is very difficult to reproduce them. The previous history of temperature variation, for example (probably critical in the establishment of corrosive film-forming organisms), is usually not known, and various parameters which existed in large field systems may be impossible to reproduce due to the extremely large number of them, some of which may only occur at the original corrosion site. The investigator always faces the possibility of having corrosion occur in the laboratory under conditions quite dissimilar (and with dissimilar mechanisms) to those which he is attempting to study. For example, in the study of the aerobic corrosion of mild steel, the purely oxidative mechanism may override any corrosive effect of microorganisms. Even where the corrosive effects of microorganisms are well established (SRB versus mild steel and *C. resinae* versus aluminum), the predominant mechanisms have not been fully established.

The microfouling aspects of the problem are now being recognized along with the increased difficulty of finding inhibitory agents which will penetrate the biofilms in relatively low concentration.

More progress in the field of biological corrosion is expected to be made since a greater number of persons have entered the field, resulting in an increase in the number of publications and conferences. This information is critical in the design and maintenance of equipment, and most importantly, in recognition of microbial corrosion problems when they occur.

Acknowledgment

I wish to thank Dr. Gregory Olson, National Bureau of Standards, for his critical review of the paper.

REFERENCES

Ainsworth, R. G., Ridgway, J., and Gwilliam, R. D. (1978). *Conf. Water Distrib. Syst. Maintenance Water Quality Pipeline Integrity* Pap. 8, pp. 1–18 (Oxford).
Alanis, I., Berado, L., DeCristofaro, N., Moina, C., and Valentini, C. (1986). *Int. Conf. Biol. Induced Corros.* pp. 102–108.
Al'bitskaya, O. N., and Shaposhnikova, N. A. (1960). *Mikrobiol. (USSR)* **29**, 725–730.
Allen, L. A., Cairns, A., Eden, G. E., Wheatland, A. B., Wormwell, F., and Nurse, T. J. (1948). *J. Soc. Chem. Ind.* **67**, 70–77.
Allred, R. C., Sudbury, J. D., and Olson, D. C. (1959). *World Oil*, 11–112.
Andreyuk, Y. I., and Kozlova, I. A. (1981). *Microbiol. Zh.* **43**, 1–9.
Argentine–U.S.A. Workshop on Biodeterioration (CONICET–NSF) (1986). Aquatec Quimica, São Paulo, Brazil.
Arnaudi, C., and Banfi, J. (1952). *Ann. Microbiol.* **5**, 26–40
Arnaudi, C., and Banfi, J. (1954). *Ann. Microbiol.* **6**, 18–40.
Banfi, J. (1952). *Chim. Ind.* **34**, 17–21.
Baru, R. L., Zinevich, A. M., Mogil'nitskii, G. M., Saposhnikova, G. A., Strarosvetskaya, Zh. O., and Timonin, V.A. (1982). *Zashch. Met.* **18**, 761-762.
Bengough, G. D., and May, R. (1924). *J. Inst. Met.* **32**, 81–269.
Bergey's Manual of Systematic Bacteriology (1984). Vol. I. Williams & Wilkins, Baltimore.
Bessems, E. (1983). *In* "Microbial Corrosion," pp. 84–89. Metals Society, London.
Bibb, M. (1986). *Int. Conf. Biol. Induced Corros.* pp. 96–101.
Biodeterioration Center (1981). Specialized Bibliography—Microbial Corrosion of Metals." Commonwealth Mycological Institute, Surrey, England.
Blanchard, G. C., and Goucher, C. R. (1967). *Electrochem. Technol.* **5**, 79–83.
Booth, G. H. (1964a). *J. Appl. Bacteriol.* **27**, 174–181.
Booth, G. H. (1964b). *Discovery* **May**, 24–27.
Booth, G. H. (1971). "Microbial Corrosion." M&B Monograph CE/I. Mills & Boon, London.
Booth, G. H., and Tiller, A. K. (1968). *Corros. Sci.* **8**, 583–600.
Booth, G. H., and Wormwell, F. (1962). *Proc. Int. Congr. Met. Corros.*, 1st pp. 341–344.
Booth. G. H., Shinn, P. M., and Wakerley, D. J. (1965). *C. R. Congr. Int. Corros. Mar. Salissures* pp. 542–554.
Booth, G. H., Cooper, A. W., and Cooper, P. M. (1967a). *Br. Corros. J.* **2**, 109–115.
Booth, G. H., Cooper, A. W., and Wakerly, D. S. (1967b). *Br. Corros. J.* **2**, 104–108.
Booth, G. H., Robb, J. A., and Wakerly, D. S. (1967c). *Proc. Int. Congr. Met. Corros.*, 3rd **2**, 542–554.
Booth, G. H., Elford, L., and Wakerly, D. S. (1968). *Br. Corros. J.* **3**, 242–245.
Bos, P., and Kuenen, J. G. (1983). *In* "Microbial Corrosion," pp. 8–27. Metals Society, London.
Burns, J. M., Staffeldt, E. E., and Calderon, O. H. (1967). *Dev. Ind. Microbiol.* **8**, 327–334.
Butlin. K. R., Vernon, W.H.J., and Whiskin, L. C. (1952). *Natl. Sanit. Eng.* **2**, 468–472.
Chantereau, J. (1981). "Corrosion Bacterienne: 2. Techniques et Documentation." 11 rue Lavoisier, Paris.
Characklis, W. G., and Cooksey, K. E. (1983). *Adv. Appl. Microbiol.* **29**, 93–138.
Churchill, A. V. (1963). *Mater. Prot. Perform.* **2**, 18–20, 22, 23.
Coles, E. L., and Davies, R. L. (1956). *Chem. Ind.* **30**, 1030–1035.
Coles, E. L., Gibson, J. G., and Hinde, R. M. (1958). *J. Appl. Chem.* **8**, 341–348.
Copenhagen, W. J. (1950). *Met. Ind.* **77**, 137.

Copenhagen, W. J. (1966). *Br. Corros. J.* **1**, 344.
Costello, J. A. (1969). *Int. Biodeterior. Bull.* **5**, 101–118.
Costello, J. A. (1974). *South Afr. J. Sci.* **70**, 202–204.
Costerton, J. W. (1983). *Dev. Ind. Microbiol.* **25**, 363–372.
Costerton, J. W., and Gessey, G. G. (1979). In "Surface Contamination" (K. L. Wittal, ed.), pp. 211–221. Plenum, New York.
Costerton, J. W., Irvin, R. T., and Cheng, K. J. (1981). *Annu. Rev. Microbiol.* **35**, 299–324.
Cragnolino, G., and MacDonald, D. D. (1982). *Corrosion* **38**, 406–424.
Cragnolino, G., and Tuovinen, O. H. (1984). *Int. Biodeterior. Bull.* **20**, 9–26.
Das, C. R., and Mishra, K. G. (1986). *Int. Conf. Biol. Induced Corros.* pp. 114–117.
Denison, I. A., and Romanoff, M. (1950). *J. Res. Natl. Bur. Stand.* **44**, 259–289.
Diekert, G., and Ritter, M. (1982). *J. Bacteriol.* **151**, 1043–1045.
Edyvean, R. G. J., and Terry, L. A. (1983). In "Biodeterioration 5" (T. A. Oxley and S. Barry, eds.), pp. 336–347. Wiley, New York.
Efird, K. D., and Lee, T. S. (1979). *Corrosion* **35**, 79–83.
Ehlert, von, I. (1967). *Mater. Org.* **2**, 297–318.
Ehlert, von, I. (1970). *Mater. Org.* **5**, 119–128.
Ehrlich, H. L. (1981). "Geomicrobiology." Dekker, New York.
Eidsa, G., and Risberg, E. (1986). *Int. Conf. Biol. Induced Corros.* pp. 109–113.
Eiselstein, L. E., Syrett, B. C., Wing, S. S., and Caligiuri, R. D. (1983). *Corros. Sci.* **23**, 223–239.
Ellis, D. (1919). "Iron Bacteria." Methuen, London.
Engel, W. B. (1969). The Role of Metallic Ion Concentrating Microorganisms in the Corrosion of Metals. *Annu. Conf., Natl. Assoc. Corros. Eng., 25th, Houston*, Pap. 36.
Evans, U. R. (1960). "Corrosion and Oxidation of Metals." Arnold, London.
Farrer, T. W., and Wormwell, F. (1953). *Chem. Ind.* **5**, 106–107.
Fischer, K. P. (1981). *Mater. Perform.* **20**, 41–46.
Frenzel, von, H. J. (1965). *Mater. Org.* **1**, 75–80.
Frenzel, von, H. J. (1966). *Mater. Org.* **1**, 275–286.
Gaines, R. H. (1910). *J. Eng. Ind. Chem.* **2**, 128–130.
Galvele, J. R. (1977). *Proc. Int. Symp. Positivity, 4th* pp. 285–327.
Garg, G. N., Sanyal, B., and Pandey, G. N. (1978). *Proc. Int. Biodeterior. Symp, 4th* pp. 99–106.
Garrett, J. H. (1891). "The Action of Water on Lead." Lewis, London.
Gerchakov, S. M., and Udey, L. R. (1982). In "Marine Biodeterioration, an Interdisciplinary Study" (J. D. Costlow and R. C. Tipper, eds.), pp. 82–87. Naval Inst. Press, Annapolis, Maryland.
Gilbert, P. T. (1946). *J. Inst. Met.* **73**, 139–174.
Grant, R., Bate, E., and Meyers, W. H. (1921). *Commonw. Eng. (Australia)* **8**, 364–366.
Greathouse, G. A., and Wessel, C. J. (1954). "Deterioration of Materials." Reinhold, New York.
Guillaume, I., Grimaudeau, G., Valensi, G., and Brison, J. (1970). *Electrochem. Acta* **15**, 1803–1825.
Guillaume, I., Grimaudeau, J., and Valensi, G. (1973a). *Corros. Sci.* **13**, 85–96.
Guillaume, I., Grimaudeau, J., and Valensi, G. (1973b). *Corros. Sci.* **13**, 97–103.
Hadley, R. F. (1948). In "The Corrosion Handbook" (H. H. Uhlig, ed.), pp. 466–481. Wiley, New York.
Hamilton, W. A. (1985). *Annu. Rev. Microbiol.* **35**, 195–217.
Hamilton, W. A., and Sanders, P. F. (1986). In "Biodeterioration 6" (S. Barry and D. R. Houghton, eds.), pp. 202–206. C.A.B. International Mycological Institute, Slough, England.

Harder, E. C. (1919). "Iron Depositing Bacteria and Their Geologic Relations." U.S. Govt. Printing Office, Washington, D.C.
Hardy, J. A. (1983). *Bro. Corros. J.* **18**, 190–193.
Harris, J. O. (1960). *Corrosion* **16**, 149T–154T.
Hedrick, H. G. (1970). *Mater Prot. Perform.* **9**, 27–31.
Hedrick, H. G., Miller, C. E., Halkias, J. E., and Hildebrand, J. F. (1964). *Appl. Microbiol.* **12**, 197–200.
Herbert, B. N., and Stott, F. D. J. (1983). In "Microbial Corrosion," pp. 7–17. Metals Society, London.
Hill, E. C. (1983). In "Microbial Corrosion," pp. 123–127. The Metals Society, London.
Hill, E. C. (1984). In "Petroleum Microbiolgy" (R. M. Atlas, ed.), pp. 579–617. Macmillan, New York.
Hill, E. C., and Thomas, A. R. (1975). *Proc. Int. Biodegrad. Symp.*, 3rd pp. 151–174. Applied Science Publ., London.
Honneysett, D. G., Bergh, van den, W. D., and O'Brien, P. F. (1985). *Mater. Perform.* **24**, 34–39.
Hutchinson, M., and Ridgway (1977). In "Aquatic Microbiology" (F. A. Skinner and J. M. Shenan, eds.), pp. 179–221. Academic Press, New York.
Iverson, W. P. (1966). *Science* **151**, 986–988.
Iverson, W. P. (1967). *Electrochem. Technol.* **5**, 77–79.
Iverson, W. P. (1968). *Nature (London)* **217**, 1265–1267.
Iverson, W. P. (1974). In "Microbial Iron Metabolism" (J. B. Neilands, ed.), pp. 475–514. Academic Press, New York.
Iverson, W. P. (1981). In "Underground Corrosion, ASTM STD 741" (E. Escalante, ed.), pp. 33–52. Am. Soc. Testing and Materials, Philadelphia.
Iverson, W. P., and Olson G. J. (1983). In "Microbial Corrosion," pp. 46–53. Metals Society, London.
Iverson, W. P., and Olson, G. J. (1984). In "Petroleum Microbiology" (R. M. Atlas, ed.), pp. 619–641. Macmillan, New York.
Iverson, W. P., Olson, G. J., and Heverly, L. F. (1986). *Int. Conf. Biol. Induced Corros.* pp. 154–161.
King, R. A. (1980). *Mater. Perform.* **19**, 39–43.
King, R. A., and Miller, J. D. A. (1971). *Nature (London)* **233**, 491–492.
King, R. A., Miller, J. D. A., and Smith, J. S. (1973a). *Br. Corros. J.* **8**, 137–141.
King, R. A., Miller, J. D. A., and Wakerly, D. S. (1973b). *Br. Corros. J.* **8**, 89–93.
King, R. A., Miller, J. D. A. and Stott, J. F. D. (1986a). *Int. Conf. Biol. Induced Corros.* pp. 268–274.
King, R. A., Skerry, B. S., Moore, D. C. A., Stott, J. F. D., and Dawson, J. L. (1986b). *Int. Conf. Biol. Induced Corros.* pp. 83–91.
Klemme, D. E., and Leonard, J. M. (1971). Naval Research Lab. Memorandum Rep. 2324. Naval Res. Lab. Washington, D.C.
Kobrin, G. (1976). *Mater. Perform.* **15**, 38–43.
Kucera, V. (1980). "Microbiological Corrosion: A Literature Survey." Swedish Corrosion Inst., Stockholm.
Kühr, von Wolzogen, C. A. H., and Vlugt, van der, L. S., (1934). *Water (Holland)* **18**, 147–165.
Lee, R. W. E. (1963). "Bibliography on Microbial Corrosion of Metals." Prevention of Deterioration Center, Division of Chemistry and Chemical Technology. Natl. Acad. Sci.–Natl. Res. Council, Washington, D.C.
Leidheiser, H., Jr. (1971). "The Corrosion of Copper, Tin and Their Alloys." Wiley, New York.

LeRoux, N. W., and Wakerley, D. S. (1978). "Microbial Corrosion. A Preliminary Survey of the Problem in U.K. Industry." CR 1505 (ME). Warren Spring Laboratory, England.
Little, B., Wagner, P., and Gerchakov, S. M. (1986). *Int. Conf. Biol. Induced Corros.* pp. 209-214.
Lutey, R. (1985). Biological Corrosion Committee. Natl. Assoc. Corros. Eng., Houston, Texas.
McCoy, J. W. (1974). "The Chemical Treatment of Cooling Water." Chem. Publ. Co., New York.
MacDonald, D. D., Roberts, B. A., and Hyne, J. B. (1978). *Corros. Sci.* **18**, 411-425.
MacDonald, D. D., Syrett, B. C., and Wing, S. S. (1979). *Corrosion* **35**, 367-378.
McDougall, J. (1969). *Aust. Corros. Eng.* **13**, 13-16.
McKenzie, P., Akbar, A. S., and Miller, J. D. A. (1977). *In* "Microbial Corrosion Affecting the Oil Industry," pp. 37-50. Institute of Petroleum, London.
Maldonado, S. B., and Boden, P. J. (1981). *Proc. Int. Congr. Met. Corros., 8th* pp. 338-343.
Mara, D. D., and Williams, D. J. A. (1971). *Corros. Sci.* **11**, 895-900.
Mara, D. D., and Williams, D. J. A. (1972a). *Corros. Sci.* **12**, 29-34.
Mara, D. D., and Williams, D. J. A. (1972b). *Br. Corros. J.* **7**, 139-142.
Metals Society (1983). "Microbial Corrosion—A Select Bibliography." Metals Society, London.
Mele, de, M. F. L., Salvarezza, R. C., and Videla, H. A. (1979). *Int. Biodeterior. Bull.* **15**, 39-44.
Miller, J. D. A. (1970). "Microbial Aspects of Metallurgy." Elsevier, New York.
Miller, J. D. A. (1981). *Microb. Biodeterior. Econ. Microbiol.* **6** 149-202.
Miller, R. N., Herron, W. C., Kregrens, A. U., Cameron, U. L., and Terry, B. M. (1964a). *Mater. Protect.* **3**, 60-67.
Miller, R. N., Herron, W. C., Krigrens, A. G., Cameron, J. L., and Terry, B. M. (1964b). *Mater. Protect.* **3**, 60-67.
Moreau, R., and Brisou, J. (1972). *Mem. Sci. Rev. Metallu.* **69**, 845-852.
Morgan, T. D. B., Steele, A. D., and Gilbert, P. D. (1983). *In* "Microbial Corrosion," pp. 66-73. Metals Society, London.
Muraoka, J. S. (1968). *Machine Design* **40**, 184-187.
National Association of Corrosion Engineers (1972). "The Role of Bacteria in the Corrosion of Oil Field Equipment." TPC. Publication No. 3. Nat. Assoc. Corros. Eng., Houston, Texas.
National Bureau of Standards (1978). "Economic Effects on Metallic Corrosion in the United States." NBS Special Pub. 511-1. Nat. Bur. Stds., Washington, D.C.
Obuekwe, C. O., Westlake, D. W. S., Cook, F. D., and Costerton, J. W. (1981a). *Appl. Environ. Microbiol.* **41**, 766-774.
Obuekwe, C. O., Westlake, D. W. S., Cook, F. D., and Plambeck, J. A. (1981b). *Corrosion* **37**, 461-467.
Olsen, E., and Szybalski, W. (1949). *Acta Chem. Scand.* **3**, 1094-1105.
OTEC (Ocean Thermal Energy Coversion) (1979). "Biofouling and Corrosion Symposium." U.S. Dept. of Commerce, Springfield, Virginia.
Pankhurst, E. S. (1968). *Appl. Bacteriol.* **31**, 179-193.
Parbery, D. G. (1971). *Mater. Org.* **6**, 161-288.
Patenaude, R. (1986). *Int. Conf. Biol. Induced Corros.* pp. 92-95.
Patterson, W. S. (1951). *Trans. Northeast Coast Eng. Ship Builders* **68**, 93-106.
Peck, H. H., and Guest, H. (1956). *J. Bacteriol.* **71**, 70-80.
Pope, D. H. (1982). Methods of Detecting, Enumerating and Determining Viability of Microorganisms Involved in Biologically Induced Corrosion. *Nat. Assoc. Corros. Eng. Corrosion/82*, Pap. No. 23.

Pope, D. H., Soracco, R. J., and Wilde, E. W. (1982). *Mater. Perform.* **7**, 43–50.
Pope, D. H., Duquette, D., Wayner, Jr., P. C., and Johannes, A. H. (1984). "Microbiologically Influenced Corrosion: A State of the Art Review." MTI Pub. No. 13. Matls. Technol. Inst. of the Chem. Proc. Indust., Columbus, Ohio.
Postgate, J. R. (1984). "The Sulphate-Reducing Bacteria," 2nd Ed. Cambridge Univ. Press, London.
Prescott, G. W. (1968). "The Algae: A Review." Houghton Mifflin, Boston.
Purkiss, B. E. (1971). In "Microbial Aspects of Metallurgy" (J. P. A. Miller, ed.), pp. 107–128. Elsevier, New York.
Ray, R. (1982). *Oil Gas J.* **80**, 87–89.
Rogers, T. H. (1948). *J. Inst. Met.* **75**, 19–38.
Romanoff, M. (1957). "Underground Corrosion." Nat. Bur. Stand. Circular 579. NBS, Washington, D.C.
Rowlands, J. C. (1965). *J. Appl. Chem.* **15**, 57–63.
Saleh, A. M., MacPherson, R., and Miller, J. D. A. (1964). *J. Appl. Bacteriol.* **27**, 281–293.
Salvarezza, R. C., Mele, de, M. F. L., and Videla, H. A. (1979). *Int. Biodeterior. Bull.* **15**, 125–132.
Salvarezza, R. C., Mele, de, M. F. L., and Videla, H. A. (1983). *Corrosion* **39**, 26–32.
Schaschl, E. (1980). *Mater. Perform.* **19**, 9–12.
Schiffrin, D. J., and Sanchez, S. R. (1985). *Corrosion* **41**, 31–38.
Schönheit, P., Moll, J., and Thauer, R. K. (1979). *Arch. Microbiol.* **123**, 105–107.
Science News (1985). **128**, 41.
Schwerdtfeger, W. J. (1967). *IEEE Trans. Ind. Gen. Appl.* **IGA-3**, 66–69.
Shrier, L. L. (1976). "Corrosion," Vol. 1. Butterworths, London.
Silva, A. J. N., Tanis, J. N., Silva, J. O., and Silva, R. A. (1986). *Int. Conf. Biol. Induced Corros.* pp. 76–82.
Smith, C. A., and Compton, K. G. (1973). *Corros. Sci.* **13**, 677–685.
Smith, W. (1968). *Cast Iron Pipe News* **May/June**, 16–29.
Southwell, C. R., and Bultman, J. D. (1972). *Proc. Inter-naval Conf. 4th, Marine Corros.* pp. 68–82.
Staffeldt, E. E., and Calderon, O. H. (1967). *Dev. Ind. Microbiol.* **8**, 321–326.
Starkey, R. L., and Wight, K. M. (1945). "Anaerobic Corrosion of Iron in Soil." Amer. Gas Assoc., New York.
Stephenson, M., and Stickland, L. H. (1931). *Biochem. J.* **25**, 215–220.
Stranger-Johannessen, M. (1986). In "Biodeterioration 6" (S. Barry and D. R. Houghton, eds.), pp. 218–223. C.A.B. International Mycological Institute, Slough, England.
Stratfull, F. F. (1961). *Corrosion* **17**, 493–496T.
Syrett, B. C., MacDonald, D. D., and Wing, S. S. (1979). *Corrosion* **35**, 409–422.
Tatnall, R. E. (1981a). *Mater. Perform.* **20**, 41–48.
Tatnall, R. E. (1981b). *Mater. Perform.* **20**, 32–38.
Taylor, J., and Parkes, R. J. (1983). *J. Gen. Microbiol.* **19**, 3303–3309.
Thorpe, P. H. (1976). *Corros. Australas* **4**, 12–14.
Tiller, A. K. (1982). In "Corrosion Processes" (R. N. Parkins, ed.), pp. 115–159. Applied Science Publ., New York.
Tiller, A. K. (1983a). In "Microbial Corrosion," pp. 104–107. Metals Society, London.
Tiller, A. K., and Booth, H. (1968). *Corros. Sci.* **8**, 549–555.
Tuovinen, O. H., Button, K. S., Vuorinen, A., Carlson, L., Mair, D. J., and Yut, L. A. (1980). *J. Am. Water Works Assoc.* **72**, 626–635.
Tuttle, R. N., and Kane, R. D., (eds.) (1981). "H_2S Corrosion in Oil and Gas Production—A Compilation of Classic Papers." Nat. Assoc. Corros. Eng., Houston, Texas.

Uhlig, H. H. (1963). "Corrosion and Corrosion Control." Wiley, New York.
Umbreit, W. (1976). *Dev. Ind. Microbiol.* **17**, 265–268.
Videla, H. A. (1981). "Corrosão Microbiologica." Biotechnologia, Vol. 4. Blücher, São Paulo.
Videla, H. A. (1986). *Int. Conf. Biol. Induced Corros.* pp. 215–222.
Videla, H. A., and Salvarezza, R. C. (1984). "Introduccion a la Corrosion Microbiologica." Mosaico Libreria Agropecuaria, Buenos Aires.
Wakerley, D. S. (1979). *Chem. Ind.* **19**, 657–659.
Walch, M., and Mitchell, R. (1986). *Int. Conf. Biol. Induced Corros.* pp. 201–208.
White, D. C. (1983). *Symp. Soc. Gen. Microbiol.* **34**, 37–66.
White, D. C., Nivens, D. E., Nichols, P. D., Mikell, A. T., Jr., Kerger, B. D., Henson, J. M., Geesey, G. G., and Clarke, C. K. (1986). *Int. Conf. Biol. Induced Corros.* pp. 233–243.
Whitney, W. R. (1903). *J. Am. Chem. Soc.* **22**, 394–406.
Widdel, F., and Pfennig, N. (1977). *Arch. Microbiol.* **112**, 119–122.
Wilkinson, T. G. (1983). *In* "Microbial Corrosion," pp. 117–122. Metals Society, London.
Willingham, C. A., and Quinby, H. L. (1971). *Dev. Ind. Microbiol.* **12**, 278–284.
Worthingham, R. G., Jack, T. R., and Ward, V. (1986). *Int. Conf. Biol. Induced Corros.* pp. 339–350.

Economics of the Bioconversion of Biomass to Methane and Other Vendable Products

RUDY J. WODZINSKI,* ROBERT N. GENNARO,*
AND MICHAEL H. SCHOLLA[†]

*Department of Biological Sciences
University of Central Florida
Orlando, Florida 32816

[†]Department of Biological Sciences
Memphis State University
Memphis, Tennessee 38152

I. Introduction

When the energy shortage of the 1970s occurred, it kindled an unprecedented interest in anaerobic digestion. Basic and applied studies were undertaken on processes to bioconvert various types of energy crops or waste substrates to methane. Many of these studies had a secondary goal of utilizing the value of the methane produced to abate pollution in an economical manner.

Earlier studies on the economics of these processes indicated that the value of the methane alone did not usually yield a profit incentive sufficient to entice many investors to underwrite the construction of methane-producing anaerobic digestion facilities. It became obvious that if profitable anaerobic digestion to produce methane was to be implemented in the near future, the process would require additional vendable products, subsidies or credits for abating pollution, or a combination of all possibilities.

A number of possibilities for vendable products and credits from anaerobic digestion have been suggested. Economic studies have been performed on the possibility of compressing the biogas and separating the carbon dioxide for sale (Wodzinski and Gennaro, 1979; Hinley, 1979), refeed of digester protein from dairy and steer fermentation (Varel et al., 1977; Ashare et al., 1977), direct sale of digester solids as fertilizer (Ashare et al., 1977), upgrading digester residues with supplemental nitrogen, phosphorus, and potassium to produce "organic fertilizer" (Wodzinski and Gennaro, 1979; Sartain, 1979), and on-site use in marketing of electricity from digester-produced methane (Rodriguez, 1979). It is also obvious that at this time the economics of the system do not support the use of substrate which is expressly produced for anaerobic digestion.

Since the onset of the renewed interest in anaerobic digestion in the 1970s, many of the key economic factors that affect the profitability of methane production have changed. It is, therefore, timely to reevaluate the economics of the production of methane and other vendable products by anaerobic digestion and to attempt to identify the areas of research that might enhance the wide-scale implementation of this process.

II. Input Data

A. GENERAL

The aim of this study was to subject the sensitive cost factors (both technical and economic) which influence the profitability of anaerobic digestion yielding methane and other vendable products to the economics of energy production operative in 1985. We have attempted, whenever possible, to use process conditions which give the highest yields of vendable products and the lowest obtainable capital costs (Tables I and II) for equipment which will perform over long periods of time with minimal maintenance costs. A further constraint on the input data that we used was that the data were derived from pilot-scale studies that had been verified as applicable to consistent performance of the process over relatively long periods of time.

III. Overview of Computer Programs

A Fortran program developed by Ashare et al. (1977) at Dynatech was modified extensively to reflect current data, credits for various potential by-products, and third-party financing (Brenner, 1983) that encompasses present (1985) tax law. The program has been revised a number of times since 1979 to encompass the advances made in rates of methane digestion and in hardware. The program as it is currently constituted retains the format established by Ashare et al. (1977) and calculates reactor performance; mass balances for liquids, solids, and gases; energy balances and needs; capital costs; and manufacturing costs.

Essentially the modified program calculates the after-tax benefits (ATB) and the costs of methane production which would accrue if an anaerobic digestion system were implemented to digest dairy, beef, swine, or poultry wastes. It has been further modified to calculate the after-tax benefits and methane value if water hyacinths (*Europa crassipes*) are digested. The program allows for various scenarios of credits to be taken, such as for carbon dioxide, fertilizer, feed, wastewater treatment, cogeneration of electricity, and use of "waste heat." The Hooke–Jeeves pattern-move portion of the Ashare et al.

TABLE I

INPUT BASELINE VARIABLES COMMON TO DAIRY, BEEF FEEDLOT, SWINE, POULTRY, AND HYACINTH PROGRAMS

Number of operators	2.5
Salary ($/hour)	7.5
Water cost ($/$10^3$ gallons)	0.40
Steam cost ($/$10^6$ BTU)	4.5
Electricity cost ($/kW-hr)	0.0752
Digester cost factor	0.0636
Vacuum filter cost factor	0.3206
Gas compressor cost factor	0.1955
Heat exchanger cost factor	3.07
Instrument cost factor	41.28
Generator cost factor	7.64
Site preparation cost factor	24.25
Maintenance–supply factor	758.5
Fraction interest debt	0.10
Fuel escalation rate	0.10
Inflation rate	0.05
Tax rate	0.46
Life of project (years)	10
Federal credits	0.20
Cost of urea ($/g)	5.27×10^{-4}
Cost of isobutylenediurea ($/g)	1.29×10^{-3}
Cost of P_2O_5 ($/g)	9.0×10^{-5}
Cost of K_2O ($/g)	9.4×10^{-5}
Desired N in fertilizer (g/kg)	60
Desired P in fertilizer (g/kg)	60
Desired K in fertilizer (g/kg)	60
Wholesale price of CO_2 ($/ton)	11
Gram fraction CO_2 (48–60°C)	0.043
Gram fraction CO_2 (25–42°C)	0.30
Gram fraction CH_4 (48–60°C)	0.57
Gram fraction CH_4 (25–42°C)	0.70
Electricity value to power grid ($/kW-hr)	0.0366
Number of fertilizer plant operators	1
Number of fertilizer plant chemists	1
Salary of fertilizer plant operator ($/hour)	7.5
Salary of fertilizer plant chemist ($/hour)	12.5
Insurance for fertilizer plant ($)	16,547.00
Digester plastic bag and pump cost factor	24.58
Pollution factor for nitrogen removal (tons/year)	8.66×10^{-3}
Pollution factor for phosphorus removal (tons/year)	2.96×10^{-3}
Wholesale value of CH_4 ($/$10^3$ ft^3)	3.00

TABLE II

INPUT BASELINE VARIABLES

Variable	Dairy	Beef	Swine	Poultry	Hyacinths
Raw materials cost factor	31.38	26.46	1×10^{-4}	0	31.38
NH_3–N in effluent (mM)	166	166	210	166	166
NH_3–N in liquid effluent (mM)	86	86	105	86	86
P in effluent (mM)	25.7	25.7	0.1	25.7	25.7
P in liquid effluent (mM)	0	0	0	0	0
K in effluent (mM)	52	0	0.1	0	52
K in liquid effluent (mM)	0	0	0	0	0
Fraction N in digested solids	0.066	0.066	0.08	0.066	0.066
Fraction P in digested solids	0.0257	0.0257	0.01	0.03	0.0257
Fraction K in digested solids	0.0563	0.0563	0.01	0.03	0.0563
Substrate cost ($/g)	2.76×10^{-6}	2.76×10^{-6}	2.76×10^{-6}	2.76×10^{-6}	0

(1977) program was modified to allow for the calculation of least-cost economics via a brute force iteration technique. An overview of the program is in Table III.

This technique was used to perform cost sensitivity analyses and determines the effects of various process parameters if the reaction rate constant is increased and if the refractory volatile solids of dairy wastes are varied. The purpose of these particular studies was to ascertain the potential value of research to increase the rate of methane production and/or the breakdown of refractory solids in dairy wastes.

A reproduction of one of the computer printouts of the program used for water hyacinths is in Table IV. In this example a conventional digester was used and a centrifuge was used to concentrate solids. None of the heat from the effluent was recovered and the gas was used internally. The printout illustrates the data obtained on a 10-acre water hyacinth pond. It includes the feed makeup process conditions, heat balance, power requirements, capital costs, manufacturing costs, including credits for fertilizer and for

TABLE III

OVERVIEW OF LOGIC FLOW FOR COMPUTER PROGRAM

I. Read in constants and options
II. Determine reaction rate constants
III. Determine reactor performance (brute force iteration) for plug flow reactor or continuous stirred reactor
IV. Determine gas production
V. Determine material mass balance
    A. Flow rate
    B. Total solids
    C. Substrate
    D. Water flow rate
    E. Nonvolatile solids
    F. Outlet total solids
        1. Centrifuge (mass of centrifuge solids)
           or
        2. No centrifuge
    G. Bleed stream
    H. Recycle
    I. Makeup water
    J. Digester volume
    K. Annual gas production
VI. Calculation of heat exchange area
VII. Determine energy balances
    A. Heat loss of digester
    B. Heat of reaction
    C. Heat loss due to water evaporation
    D. Heat loss due to bleed stream (liquid and solids)
        1. With heat exchanger
        2. Without heat exchanger
    E. Heat loss with gas
    F. Heat loss due to water makeup
    G. Heat to operate fertilizer plant
        1. With fertilizer plant: Excess heat from electrical generator to dry fertilizer
        2. No fertilizer plant
    H. Heat loss from slurry
        1. With heat exchanger
        2. Without heat exchanger
    I. Substrate heat loss
    J. Determine horsepower requirements
        1. Horsepower requirement for mixer or slurry
        2. Horsepower requirement for digester if mixed
        3. Horsepower requirement for centrifuge
        4. Horsepower subtotal
    K. Determine on-site energy uses (option)
        1. Annual amount of gas used to produce electricty
        2. Annual amount of gas used to heat water
        3. Total internal annual gas usage
        4. Excess heat produced by on-site generation of electricity

(continued)

TABLE III *(Continued)*

- L. Calculation of total heat needs
- M. Calculation of horsepower required to compress gas
- N. Calculation of total horsepower requirements

VIII. Program cost
- A. Read in variables
- B. Determine capital cost of components
    1. Digester cost
        a. Conventional, includes digester heat exchanger, insulation, and valves
        b. Plastic, includes pumps
    2. Centrifuge cost if used
    3. Gas compressor, storage tanks, meters, pressure-relief valves, and switches
    4. Sludge drying bed cost (if used)
    5. Heat exchanger cost (effluent)
    6. Fertilizer plant cost
    7. Site preparation cost
    8. Harvester, grinder, and pumps (for hyacinths only)
- C. Calculate plant investment cost
    1. Contractors fee cost
    2. Engineering fee cost
    3. Cost to confine cattle (if used)
    4. Gas compression piping (if used)
    5. Sewer pipeline for substrate transport (if used)
    6. Capital to construct ponds (hyacinths only)
    7. Calculate subtotal plant investment cost
    8. Project contingency cost
    9. Calculate total plant investment cost
- D. Calculate annual manufacturing cost
    1. Raw material cost
    2. Determine if fertilizer plant is desired and if solids production is adequate
        a. Determine amount of NPK needed for desired levels if fertilizer used
        b. Determine cost of NPK needed for desired levels if fertilizer used
        c. Determine annual water cost
        d. Determine annual steam cost
        e. Determine annual utility cost
        f. Determine annual labor cost
        g. Determine annual administrative overhead costs
        h. Determine annual supplies cost
        i. Determine annual maintenance–supply cost
        j. Determine annual taxes and insurance costs
        k. Determine operating cost for fertilizer plant (if used)
        l. Determine fertilizer plant insurance cost
        m. Calculate annual gross operating cost
        n. Calculate credits for vendable products other than CH_4
            1. Fertilizer credit (if used)
            2. CO_2 credit (if used)
            3. Electricity credit (if used)

(continued)

The two organisms most commonly found associated with aluminum corrosion, *Pseudomonas* spp. and *Cladosporium* spp., have been considered the most active species (Parbery, 1971; Hedrick et al., 1964). A strain of *Pseudomonas aeruginosa* was found by Blanchard and Goucher (1967) to produce corrosive organic compounds (large molecules with molecular weight greater than 5000). Hedrick (1970), using a mixed inoculum of *C. resinae* and *Pseudomonas* sp. with four aluminum alloys (2024, 7075, 7079, and 1100), found alloy 7075 to be most susceptible and alloy 7079 the least susceptible to corrosion. It was proposed that the corrosion resulted from the removal of zinc and magnesium from the basic structure of the alloy by extracellular enzymatic activity.

Evidence that *C. resinae* causes extensive corrosion of aluminum aircraft alloys was reported by McKenzie et al. (1977). Using pure cultures growing on Bushnell–Haas medium with kerosene, they found pitting, intergranular corrosion, and exfoliation. Corroded regions were found to be rich in copper and iron, surrounded by areas depleted in zinc, magnesium, and aluminum. Citric, isocitric, cis-aconitic, and α-oxoglutaric acids, which had accumulated during growth, were found to be quite corrosive to the aluminum when tested individually against aircraft alloys 2024–2073 and 7075–7076. The corrosion rates were found to be comparable to those in the fungus cultures. In order to elucidate the importance of *C. resinae* and *P. aeruginosa* in the corrosion of aluminum, Mele et al. (1979) and Salvarezza et al. (1979) made studies of the influence of these organisms on the pH and the pitting potential (E_p) of aluminum. The pitting potential of a metal may be defined in a potentiostatic polarization curve as the potential below which the metal surface remains passive and above which pitting nucleates the metal surface. The increase in the concentration of an aggressive substance in a medium produces a decrease in the E_p value (Galvele, 1977). It was found that a marked decrease of pH and the pitting potential occurred in cultures of *C. resinae* but not in cultures of *P. aeruginosa*. It was concluded that *C. resinae* plays a more relevant role in aluminum corrosion by the production of corrosive metabolic products. As it now appears, the role of metabolites from *C. resinae* appears to be quite well established, but corrosion by other organisms and mechanisms cannot be completely ruled out.

C. Other Metals

Although not investigated thoroughly, there is evidence that zinc and lead undergo rapid corrosion under conditions favorable to the growth of SRB (Denison and Romanoff, 1950). Corrosion of zinc by SRB has been reported by Arnaudi and Banfi (1952, 1954), Banfi (1952), and Garg et al. (1978). Banfi (1952) found that the corrosion of zinc by these bacteria was localized in the zone where the air–liquid interface and hydrogen sulfide were in contact with the metal, indicating that the corrosion probably was due to acids produced by the oxidation of sulfide. Zinc would therefore appear to be corroded by any acid-producing microorganism, as Ehlert (1970) and Staffeldt and Calderon (1967) have demonstrated. The corrosion of zinc was found by Ehlert to be less, however, than that of iron due to the inhibitory effect of the zinc ions on bacterial development.

The corrosion of lead underground cable sheath known as "phenol corrosion" was demonstrated by Coles et al. (1956, 1958) to be produced by low-molecular-weight organic acids from the bacterial decomposition of cellulosic wrappings. A number of organic acids produced by microorganisms have been demonstrated to be corrosive to lead (Staffeldt and Calderon, 1967). Romanoff (1957) concluded that in soil conditions associated with deficiency of oxygen and high organic acidity (pH 4–5), lead is corroded. Since sulfate-rich soils were inhibitory to the corrosion and SRB are not very acid tolerant, they probably did not play a significant role in this type of corrosion.

The effect of 17 types of bacteria on the corrosion of nickel in fresh water and seawater was investigated by Guillaume et al. (1973b), who found that severe corrosion occurred upon disappearance of the bacteria, which they ascribed to a high pH, a high redox potential, and dissolved nickel ions. A nickel-, cobalt-, and molybdenum-dependent strain of Methanobacterium thermoautotrophicum has been demonstrated (Schönheit et al., 1979) as well as a requirement by various clostridia and Acetobacterium woodii for nickel in the reduction of CO_2 to acetate as part of their energy metabolism (Diekert and Ritter, 1982). It would be of interest to determine their action on metallic nickel and its alloys. A suspected case of microbial corrosion of nickel-based alloys (Ni–Cu and Ni–Mo) has been reported (Kobrin, 1976).

The corrosion of high-purity magnesium as indicated by weight increase of P. aeruginosa, Bacillus sp., and C. resinae was reported by Hedrick (1970). The corrosion test medium consisted of saline-washed inoculum, deionized water, and a JP-4 (kerosene type) fuel overlay. It was, however, not indicated whether the control test medium contained saline solution.

No instances of microbiological corrosion of titanium have been reported (Pope et al., 1984).

VII. Prevention and Control

A. Selection and Control of Environment

Whenever possible, the environment in which metals are to be used should be assessed for corrosivity. Selection of a less corrosive environment would thereby alleviate later corrosion and corrosion control problems. Also, if metals are already in contact with soil, for example, assessment of the soil as corrosive would enable the proper protective measures to be taken.

In the case of soil in which pipelines are to be laid, a number of factors have been considered by various authors as indicative of corrosivity. These include redox potential (Starkey and Wright, 1945), soil resistivity (Romanoff, 1957), pH, and water content (Booth et al., 1967b). Stratful (1961) has indicated an optimum water content to give a minimum resistivity that can be correlated with corrosion aggressiveness. In the United Kingdom, a soil-testing scheme, reported by Booth et al. (1967a,b), to indicate the aggressiveness of soils to ferrous metals, involves the determination of three criteria: (1) soil resistivity, (2) redox potential, and (3) water content. In the United States, a scheme proposed by the Cast Iron Pipe Research Association (Smith, 1968) has involved five criteria: (1) soil resistivity, (2) pH, (3) redox potential, (4) sulfide content, and (5) moisture content. According to a survey in the United Kingdom, soil resistivity was the most favored single test used (LeRoux and Wakerly, 1978).

With an increase in oil-producing activity in the North Sea, a scheme to test the corrosivity of seabeds has been devised by King (1980). This scheme includes five criteria: (1) type of sediment, (2) organic content, (3) water depth, (4) nitrogen and phosphorus content, and (5) temperature.

Some idea of the corrosiveness of water may be ascertained from (1) total number of bacteria, (2) types and number of bacteria, (3) redox potential, (4) temperature, (5) salinity, (6) concentration of organic material, and (7) concentration of sulfide and ammonium ions in addition to other dissolved ions. Probably the most reliable method for testing soil and water aggressivity is to use test specimens, if time is available, or electrochemical corrosion rate methods, if a quick assessment is desired.

In most cases, modification of the environment is difficult or very costly. In the case of short sections of pipe, protection against SRB may be accomplished by avoidance of anaerobic conditions, by drainage or gravel surrounds (Butlin et al., 1952), or by using a backfill such as chalk or lime, which provides a sufficient alkaline environment to inhibit the growth of SRB (Hadley, 1948). The use of biocides in the backfill has also been suggested.

B. COATINGS

The use of protective coatings to provide a barrier between the metal and its corrosive environment is an old but effective measure of protection. Materials which have been employed include coal tar epoxies and enamels, asphaltic bitumens, epoxy resins, and various cements. In many cases, failures have occurred due to poor application or to mechanical damage from handling and backfilling.

Disbonded plastic (polythene) tapes and sleeving have been used in the United States for several years and have been considered as performing successfully. A recent report, however, has implicated SRB in pipeline leaks (corrosion) occurring beneath disbonded plastic tape coatings (Worthingham et al., 1985). The most effective corrosion protection system, however, is one employing cathodic protection of coated pipe which ensures protection of breaks (holidays) in the coating.

C. CATHODIC PROTECTION

The application of an electric current to metal counteracts natural corrosion currents, that is, supplies sufficient electrons to the metal so that the metal cations cannot escape from the metal. Two methods for accomplishing this are to use metal sacrificial anodes (aluminum, magnesium, or zinc), which corrode and thereby supply the electrons to the protected metal (in the case of iron), or to use an inert anode, electrons being supplied from a rectifier. For the protection of mild steel in the soil, a potential of -0.85 V, with respect to a $Cu-CuSO_4$ half cell, is usually maintained. If the soil is aggressive (a high risk of microbial corrosion), a potential of -0.95 V is recommended (Booth and Tiller, 1968; Fischer, 1981). Corrosion, in which SRB were implicated, of pipelines beneath disbonded plastic tape coatings was reported in which cathodic protection potentials were even above -1.00 V (Worthingham et al., 1986).

In cases where copper has been cathodically protected in a marine environment, fouling has occurred due to the prevention of the Cu ion formation that normally prevents fouling (Gerchakov and Udey, 1982). In practice, it is customary, however, to use cathodic protection on coated structures. The small protection current will be used to protect only those areas which the coating does not cover, such as small pinholes.

D. INHIBITORS

In closed systems such as tanks and recirculating cooling-water systems, the use of chemicals both to inhibit corrosion as well as to inhibit microbial growth may be employed. Discussions of anticorrosive chemicals may be found in the many books on general corrosion.

Presently, there are a number of biocides or biostats in use. These include oxidizing agents such as chlorine and ozone, phenolics, aldehydes, metal organic compounds such as organotins, heavy metal salts, and quaternary ammonium compounds.

Saleh et al. (1964) have listed the effect of some 200 biocides and bacteriostats on SRB. A list of potential compounds with microbiocidal/microbiostatic activity as well as other measures that may be taken to control biological corrosion in oil field equipment has been outlined (National Association of Corrosion Engineers, 1972). A recent assessment of some of these compounds for use in the oil recovery system is given by Bessems (1983). A review of commercial biocides for use in industrial water systems that are subject to microbial corrosion has been published (McCoy, 1974). In aircraft fuel tanks, organoborane compounds, potassium dichromate, and glycol monomethyl ether have been reported to be effective in preventing fungal corrosion (Miller, 1981). The latter compound is also an icing inhibitor which, when added to the fuel, partitions to the water phase to form inhibitory concentrations. A review of biocides for use in cutting oil and other types of oils has been published by Hill (1984).

In general, the control of microorganisms that form films that result in potential corrosion is much more difficult and requires much higher concentrations of inhibitory agents than the control of microorganisms in the aqueous phase (Costerton, 1983). Use of several compatible inhibitory compounds in rotation or simultaneously may be more successful than continued use of a single compound, since the likelihood of microbial tolerance may be decreased. The compounds should also be compatible with the corrosion inhibitor, if used. A reported case of successful corrosion control in a cooling-water system was obtained by employing an organic film corrosion inhibitor, a polyacrylate/phosphonate dispersant, and a combination of two biocides used simultaneously (Honneysett et al., 1985).

VIII. Conclusions

It is evident, from the abundance of literature on microbially induced corrosion and from various assessments of its economic effect, that microorganisms probably play a significant role in the corrosion process. This conclusion is further buttressed by good evidence that corrosion is inhibited by the use of microbial inhibitory compounds (viz., aluminum fuel tank corrosion).

Much of the evidence is suggestive, however. Relatively few reports present direct evidence, that is, evidence in accordance with Koch's postulates, namely (1) isolating the organisms from a corroded area in the field, (2) reproducing the corrosion under identical conditions in the laboratory using appropriate sterile controls, and (3) reisolating the identical organisms from the corrosion site in the laboratory. The most difficult postulate is the reproduction of corrosion under identical conditions as first noticed in the field. More often than not, the field conditions are not known and the corroded sample, usually well dehydrated, is all the evidence available. In other instances, where field conditions do exist, it is very difficult to reproduce them. The previous history of temperature variation, for example (probably critical in the establishment of corrosive film-forming organisms), is usually not known, and various parameters which existed in large field systems may be impossible to reproduce due to the extremely large number of them, some of which may only occur at the original corrosion site. The investigator always faces the possibility of having corrosion occur in the laboratory under conditions quite dissimilar (and with dissimilar mechanisms) to those which he is attempting to study. For example, in the study of the aerobic corrosion of mild steel, the purely oxidative mechanism may override any corrosive effect of microorganisms. Even where the corrosive effects of microorganisms are well established (SRB versus mild steel and *C. resinae* versus aluminum), the predominant mechanisms have not been fully established.

The microfouling aspects of the problem are now being recognized along with the increased difficulty of finding inhibitory agents which will penetrate the biofilms in relatively low concentration.

More progress in the field of biological corrosion is expected to be made since a greater number of persons have entered the field, resulting in an increase in the number of publications and conferences. This information is critical in the design and maintenance of equipment, and most importantly, in recognition of microbial corrosion problems when they occur.

Acknowledgment

I wish to thank Dr. Gregory Olson, National Bureau of Standards, for his critical review of the paper.

References

Ainsworth, R. G., Ridgway, J., and Gwilliam, R. D. (1978). *Conf. Water Distrib. Syst. Maintenance Water Quality Pipeline Integrity* Pap. 8, pp. 1–18 (Oxford).
Alanis, I., Berado, L., DeCristofaro, N., Moina, C., and Valentini, C. (1986). *Int. Conf. Biol. Induced Corros.* pp. 102–108.
Al'bitskaya, O. N., and Shaposhnikova, N. A. (1960). *Mikrobiol. (USSR)* **29**, 725–730.
Allen, L. A., Cairns, A., Eden, G. E., Wheatland, A. B., Wormwell, F., and Nurse, T. J. (1948). *J. Soc. Chem. Ind.* **67**, 70–77.
Allred, R. C., Sudbury, J. D., and Olson, D. C. (1959). *World Oil*, 11–112.
Andreyuk, Y. I., and Kozlova, I. A. (1981). *Microbiol. Zh.* **43**, 1–9.
Argentine–U.S.A. Workshop on Biodeterioration (CONICET–NSF) (1986). Aquatec Quimica, São Paulo, Brazil.
Arnaudi, C., and Banfi, J. (1952). *Ann. Microbiol.* **5**, 26–40
Arnaudi, C., and Banfi, J. (1954). *Ann. Microbiol.* **6**, 18–40.
Banfi, J. (1952). *Chim. Ind.* **34**, 17–21.
Baru, R. L., Zinevich, A. M., Mogil'nitskii, G. M., Saposhnikova, G. A., Strarosvetskaya, Zh. O., and Timonin, V.A. (1982). *Zashch. Met.* **18**, 761-762.
Bengough, G. D., and May, R. (1924). *J. Inst. Met.* **32**, 81–269.
Bergey's Manual of Systematic Bacteriology (1984). Vol. I. Williams & Wilkins, Baltimore.
Bessems, E. (1983). *In* "Microbial Corrosion," pp. 84–89. Metals Society, London.
Bibb, M. (1986). *Int. Conf. Biol. Induced Corros.* pp. 96–101.
Biodeterioration Center (1981). Specialized Bibliography—Microbial Corrosion of Metals." Commonwealth Mycological Institute, Surrey, England.
Blanchard, G. C., and Goucher, C. R. (1967). *Electrochem. Technol.* **5**, 79–83.
Booth, G. H. (1964a). *J. Appl. Bacteriol.* **27**, 174–181.
Booth, G. H. (1964b). *Discovery* **May**, 24–27.
Booth, G. H. (1971). "Microbial Corrosion." M&B Monograph CE/I. Mills & Boon, London.
Booth, G. H., and Tiller, A. K. (1968). *Corros. Sci.* **8**, 583–600.
Booth, G. H., and Wormwell, F. (1962). *Proc. Int. Congr. Met. Corros., 1st* pp. 341–344.
Booth. G. H., Shinn, P. M., and Wakerley, D. J. (1965). *C. R. Congr. Int. Corros. Mar. Salissures* pp. 542–554.
Booth, G. H., Cooper, A. W., and Cooper, P. M. (1967a). *Br. Corros. J.* **2**, 109–115.
Booth, G. H., Cooper, A. W., and Wakerly, D. S. (1967b). *Br. Corros. J.* **2**, 104–108.
Booth, G. H., Robb, J. A., and Wakerly, D. S. (1967c). *Proc. Int. Congr. Met. Corros., 3rd* **2**, 542–554.
Booth, G. H., Elford, L., and Wakerly, D. S. (1968). *Br. Corros. J.* **3**, 242–245.
Bos, P., and Kuenen, J. G. (1983). *In* "Microbial Corrosion," pp. 8–27. Metals Society, London.
Burns, J. M., Staffeldt, E. E., and Calderon, O. H. (1967). *Dev. Ind. Microbiol.* **8**, 327–334.
Butlin. K. R., Vernon, W.H.J., and Whiskin, L. C. (1952). *Natl. Sanit. Eng.* **2**, 468–472.
Chantereau, J. (1981). "Corrosion Bacterienne: 2. Techniques et Documentation." 11 rue Lavoisier, Paris.
Characklis, W. G., and Cooksey, K. E. (1983). *Adv. Appl. Microbiol.* **29**, 93–138.
Churchill, A. V. (1963). *Mater. Prot. Perform.* **2**, 18–20, 22, 23.
Coles, E. L., and Davies, R. L. (1956). *Chem. Ind.* **30**, 1030–1035.
Coles, E. L., Gibson, J. G., and Hinde, R. M. (1958). *J. Appl. Chem.* **8**, 341–348.
Copenhagen, W. J. (1950). *Met. Ind.* **77**, 137.

Copenhagen, W. J. (1966). *Br. Corros. J.* **1**, 344.
Costello, J. A. (1969). *Int. Biodeterior. Bull.* **5**, 101–118.
Costello, J. A. (1974). *South Afr. J. Sci.* **70**, 202–204.
Costerton, J. W. (1983). *Dev. Ind. Microbiol.* **25**, 363–372.
Costerton, J. W., and Gessey, G. G. (1979). *In* "Surface Contamination" (K. L. Wittal, ed.), pp. 211–221. Plenum, New York.
Costerton, J. W., Irvin, R. T., and Cheng, K. J. (1981). *Annu. Rev. Microbiol.* **35**, 299–324.
Cragnolino, G., and MacDonald, D. D. (1982). *Corrosion* **38**, 406–424.
Cragnolino, G., and Tuovinen, O. H. (1984). *Int. Biodeterior. Bull.* **20**, 9–26.
Das, C. R., and Mishra, K. G. (1986). *Int. Conf. Biol. Induced Corros.* pp. 114–117.
Denison, I. A., and Romanoff, M. (1950). *J. Res. Natl. Bur. Stand.* **44**, 259–289.
Diekert, G., and Ritter, M. (1982). *J. Bacteriol.* **151**, 1043–1045.
Edyvean, R. G. J., and Terry, L. A. (1983). *In* "Biodeterioration 5" (T. A. Oxley and S. Barry, eds.), pp. 336–347. Wiley, New York.
Efird, K. D., and Lee, T. S. (1979). *Corrosion* **35**, 79–83.
Ehlert, von, I. (1967). *Mater. Org.* **2**, 297–318.
Ehlert, von, I. (1970). *Mater. Org.* **5**, 119–128.
Ehrlich, H. L. (1981). "Geomicrobiology." Dekker, New York.
Eidsa, G., and Risberg, E. (1986). *Int. Conf. Biol. Induced Corros.* pp. 109–113.
Eiselstein, L. E., Syrett, B. C., Wing, S. S., and Caligiuri, R. D. (1983). *Corros. Sci.* **23**, 223–239.
Ellis, D. (1919). "Iron Bacteria." Methuen, London.
Engel, W. B. (1969). The Role of Metallic Ion Concentrating Microorganisms in the Corrosion of Metals. *Annu. Conf., Natl. Assoc. Corros. Eng., 25th, Houston,* Pap. 36.
Evans, U. R. (1960). "Corrosion and Oxidation of Metals." Arnold, London.
Farrer, T. W., and Wormwell, F. (1953). *Chem. Ind.* **5**, 106–107.
Fischer, K. P. (1981). *Mater. Perform.* **20**, 41–46.
Frenzel, von, H. J. (1965). *Mater. Org.* **1**, 75–80.
Frenzel, von, H. J. (1966). *Mater. Org.* **1**, 275–286.
Gaines, R. H. (1910). *J. Eng. Ind. Chem.* **2**, 128–130.
Galvele, J. R. (1977). *Proc. Int. Symp. Positivity, 4th* pp. 285–327.
Garg, G. N., Sanyal, B., and Pandey, G. N. (1978). *Proc. Int. Biodeterior. Symp, 4th* pp. 99–106.
Garrett, J. H. (1891). "The Action of Water on Lead." Lewis, London.
Gerchakov, S. M., and Udey, L. R. (1982). *In* "Marine Biodeterioration, an Interdisciplinary Study" (J. D. Costlow and R. C. Tipper, eds.), pp. 82–87. Naval Inst. Press, Annapolis, Maryland.
Gilbert, P. T. (1946). *J. Inst. Met.* **73**, 139–174.
Grant, R., Bate, E., and Meyers, W. H. (1921). *Commonw. Eng. (Australia)* **8**, 364–366.
Greathouse, G. A., and Wessel, C. J. (1954). "Deterioration of Materials." Reinhold, New York.
Guillaume, I., Grimaudeau, G., Valensi, G., and Brison, J. (1970). *Electrochem. Acta* **15**, 1803–1825.
Guillaume, I., Grimaudeau, J., and Valensi, G. (1973a). *Corros. Sci.* **13**, 85–96.
Guillaume, I., Grimaudeau, J., and Valensi, G. (1973b). *Corros. Sci.* **13**, 97–103.
Hadley, R. F. (1948). *In* "The Corrosion Handbook" (H. H. Uhlig, ed.), pp. 466–481. Wiley, New York.
Hamilton, W. A. (1985). *Annu. Rev. Microbiol.* **35**, 195–217.
Hamilton, W. A., and Sanders, P. F. (1986). *In* "Biodeterioration 6" (S. Barry and D. R. Houghton, eds.), pp. 202–206. C.A.B. International Mycological Institute, Slough, England.

Harder, E. C. (1919). "Iron Depositing Bacteria and Their Geologic Relations." U.S. Govt. Printing Office, Washington, D.C.
Hardy, J. A. (1983). *Bro. Corros. J.* **18**, 190–193.
Harris, J. O. (1960). *Corrosion* **16**, 149T-154T.
Hedrick, H. G. (1970). *Mater Prot. Perform.* **9**, 27–31.
Hedrick, H. G., Miller, C. E., Halkias, J. E., and Hildebrand, J. F. (1964). *Appl. Microbiol.* **12**, 197–200.
Herbert, B. N., and Stott, F. D. J. (1983). In "Microbial Corrosion," pp. 7–17. Metals Society, London.
Hill, E. C. (1983). In "Microbial Corrosion," pp. 123–127. The Metals Society, London.
Hill, E. C. (1984). In "Petroleum Microbiolgy" (R. M. Atlas, ed.), pp. 579–617. Macmillan, New York.
Hill, E. C., and Thomas, A. R. (1975). *Proc. Int. Biodegrad. Symp.*, 3rd pp. 151–174. Applied Science Publ., London.
Honneysett, D. G., Bergh, van den, W. D., and O'Brien, P. F. (1985). *Mater. Perform.* **24**, 34–39.
Hutchinson, M., and Ridgway (1977). In "Aquatic Microbiology" (F. A. Skinner and J. M. Shenan, eds.), pp. 179–221. Academic Press, New York.
Iverson, W. P. (1966). *Science* **151**, 986–988.
Iverson, W. P. (1967). *Electrochem. Technol.* **5**, 77–79.
Iverson, W. P. (1968). *Nature (London)* **217**, 1265–1267.
Iverson, W. P. (1974). In "Microbial Iron Metabolism" (J. B. Neilands, ed.), pp. 475–514. Academic Press, New York.
Iverson, W. P. (1981). In "Underground Corrosion, ASTM STD 741" (E. Escalante, ed.), pp. 33–52. Am. Soc. Testing and Materials, Philadelphia.
Iverson, W. P., and Olson G. J. (1983). In "Microbial Corrosion," pp. 46–53. Metals Society, London.
Iverson, W. P., and Olson, G. J. (1984). In "Petroleum Microbiology" (R. M. Atlas, ed.), pp. 619–641. Macmillan, New York.
Iverson, W. P., Olson, G. J., and Heverly, L. F. (1986). *Int. Conf. Biol. Induced Corros.* pp. 154–161.
King, R. A. (1980). *Mater. Perform.* **19**, 39–43.
King, R. A., and Miller, J. D. A. (1971). *Nature (London)* **233**, 491–492.
King, R. A., Miller, J. D. A., and Smith, J. S. (1973a). *Br. Corros. J.* **8**, 137–141.
King, R. A., Miller, J. D. A., and Wakerly, D. S. (1973b). *Br. Corros. J.* **8**, 89–93.
King, R. A., Miller, J. D. A. and Stott, J. F. D. (1986a). *Int. Conf. Biol. Induced Corros.* pp. 268–274.
King, R. A., Skerry, B. S., Moore, D. C. A., Stott, J. F. D., and Dawson, J. L. (1986b). *Int. Conf. Biol. Induced Corros.* pp. 83–91.
Klemme, D. E., and Leonard, J. M. (1971). Naval Research Lab. Memorandum Rep. 2324. Naval Res. Lab. Washington, D.C.
Kobrin, G. (1976). *Mater. Perform.* **15**, 38–43.
Kucera, V. (1980). "Microbiological Corrosion: A Literature Survey." Swedish Corrosion Inst., Stockholm.
Kühr, von Wolzogen, C. A. H., and Vlugt, van der, L. S., (1934). *Water (Holland)* **18**, 147–165.
Lee, R. W. E. (1963). "Bibliography on Microbial Corrosion of Metals." Prevention of Deterioration Center, Division of Chemistry and Chemical Technology. Natl. Acad. Sci.–Natl. Res. Council, Washington, D.C.
Leidheiser, H., Jr. (1971). "The Corrosion of Copper, Tin and Their Alloys." Wiley, New York.

LeRoux, N. W., and Wakerley, D. S. (1978). "Microbial Corrosion. A Preliminary Survey of the Problem in U.K. Industry." CR 1505 (ME). Warren Spring Laboratory, England.

Little, B., Wagner, P., and Gerchakov, S. M. (1986). *Int. Conf. Biol. Induced Corros.* pp. 209–214.

Lutey, R. (1985). Biological Corrosion Committee. Natl. Assoc. Corros. Eng., Houston, Texas.

McCoy, J. W. (1974). "The Chemical Treatment of Cooling Water." Chem. Publ. Co., New York.

MacDonald, D. D., Roberts, B. A., and Hyne, J. B. (1978). *Corros. Sci.* **18**, 411–425.

MacDonald, D. D., Syrett, B. C., and Wing, S. S. (1979). *Corrosion* **35**, 367–378.

McDougall, J. (1969). *Aust. Corros. Eng.* **13**, 13–16.

McKenzie, P., Akbar, A. S., and Miller, J. D. A. (1977). In "Microbial Corrosion Affecting the Oil Industry," pp. 37–50. Institute of Petroleum, London.

Maldonado, S. B., and Boden, P. J. (1981). *Proc. Int. Congr. Met. Corros.*, 8th pp. 338–343.

Mara, D. D., and Williams, D. J. A. (1971). *Corros. Sci.* **11**, 895–900.

Mara, D. D., and Williams, D. J. A. (1972a). *Corros. Sci.* **12**, 29–34.

Mara, D. D., and Williams, D. J. A. (1972b). *Br. Corros. J.* **7**, 139–142.

Metals Society (1983). "Microbial Corrosion—A Select Bibliography." Metals Society, London.

Mele, de, M. F. L., Salvarezza, R. C., and Videla, H. A. (1979). *Int. Biodeterior. Bull.* **15**, 39–44.

Miller, J. D. A. (1970). "Microbial Aspects of Metallurgy." Elsevier, New York.

Miller, J. D. A. (1981). *Microb. Biodeterior. Econ. Microbiol.* **6** 149–202.

Miller, R. N., Herron, W. C., Kregrens, A. U., Cameron, U. L., and Terry, B. M. (1964a). *Mater. Protect.* **3**, 60–67.

Miller, R. N., Herron, W. C., Krigrens, A. G., Cameron, J. L., and Terry, B. M. (1964b). *Mater. Protect.* **3**, 60–67.

Moreau, R., and Brisou, J. (1972). *Mem. Sci. Rev. Metallu.* **69**, 845–852.

Morgan, T. D. B., Steele, A. D., and Gilbert, P. D. (1983). In "Microbial Corrosion," pp. 66–73. Metals Society, London.

Muraoka, J. S. (1968). *Machine Design* **40**, 184–187.

National Association of Corrosion Engineers (1972). "The Role of Bacteria in the Corrosion of Oil Field Equipment." TPC. Publication No. 3. Nat. Assoc. Corros. Eng., Houston, Texas.

National Bureau of Standards (1978). "Economic Effects on Metallic Corrosion in the United States." NBS Special Pub. 511-1. Nat. Bur. Stds., Washington, D.C.

Obuekwe, C. O., Westlake, D. W. S., Cook, F. D., and Costerton, J. W. (1981a). *Appl. Environ. Microbiol.* **41**, 766–774.

Obuekwe, C. O., Westlake, D. W. S., Cook, F. D., and Plambeck, J. A. (1981b). *Corrosion* **37**, 461–467.

Olsen, E., and Szybalski, W. (1949). *Acta Chem. Scand.* **3**, 1094–1105.

OTEC (Ocean Thermal Energy Coversion) (1979). "Biofouling and Corrosion Symposium." U.S. Dept. of Commerce, Springfield, Virginia.

Pankhurst, E. S. (1968). *Appl. Bacteriol.* **31**, 179–193.

Parbery, D. G. (1971). *Mater. Org.* **6**, 161–288.

Patenaude, R. (1986). *Int. Conf. Biol. Induced Corros.* pp. 92–95.

Patterson, W. S. (1951). *Trans. Northeast Coast Eng. Ship Builders* **68**, 93–106.

Peck, H. H., and Guest, H. (1956). *J. Bacteriol.* **71**, 70–80.

Pope, D. H. (1982). Methods of Detecting, Enumerating and Determining Viability of Microorganisms Involved in Biologically Induced Corrosion. *Nat. Assoc. Corros. Eng. Corrosion/82*, Pap. No. 23.

Pope, D. H., Soracco, R. J., and Wilde, E. W. (1982). *Mater. Perform.* **7**, 43–50.
Pope, D. H., Duquette, D., Wayner, Jr., P. C., and Johnannes, A. H. (1984). "Microbiologically Influenced Corrosion: A State of the Art Review." MTI Pub. No. 13. Matls. Technol. Inst. of the Chem. Proc. Indust., Columbus, Ohio.
Postgate, J. R. (1984). "The Sulphate-Reducing Bacteria," 2nd Ed. Cambridge Univ. Press, London.
Prescott, G. W. (1968). "The Algae: A Review." Houghton Mifflin, Boston.
Purkiss, B. E. (1971). In "Microbial Aspects of Metallurgy" (J. P. A. Miller, ed.), pp. 107–128. Elsevier, New York.
Ray, R. (1982). *Oil Gas J.* **80**, 87–89.
Rogers, T. H. (1948). *J. Inst. Met.* **75**, 19–38.
Romanoff, M. (1957). "Underground Corrosion." Nat. Bur. Stand. Circular 579. NBS, Washington, D.C.
Rowlands, J. C. (1965). *J. Appl. Chem.* **15**, 57–63.
Saleh, A. M., MacPherson, R., and Miller, J. D. A. (1964). *J. Appl. Bacteriol.* **27**, 281–293.
Salvarezza, R. C., Mele, de, M. F. L., and Videla, H. A. (1979). *Int. Biodeterior. Bull.* **15**, 125–132.
Salvarezza, R. C., Mele, de, M. F. L., and Videla, H. A. (1983). *Corrosion* **39**, 26–32.
Schaschl, E. (1980). *Mater. Perform.* **19**, 9–12.
Schiffrin, D. J., and Sanchez, S. R. (1985). *Corrosion* **41**, 31–38.
Schönheit, P., Moll, J., and Thauer, R. K. (1979). *Arch. Microbiol.* **123**, 105–107.
Science News (1985). **128**, 41.
Schwerdtfeger, W. J. (1967). *IEEE Trans. Ind. Gen. Appl.* **IGA-3**, 66–69.
Shrier, L. L. (1976). "Corrosion," Vol. 1. Butterworths, London.
Silva, A. J. N., Tanis, J. N., Silva, J. O., and Silva, R. A. (1986). *Int. Conf. Biol. Induced Corros.* pp. 76–82.
Smith, C. A., and Compton, K. G. (1973). *Corros. Sci.* **13**, 677–685.
Smith, W. (1968). *Cast Iron Pipe News* **May/June**, 16–29.
Southwell, C. R., and Bultman, J. D. (1972). *Proc. Inter-naval Conf. 4th, Marine Corros.* pp. 68–82.
Staffeldt, E. E., and Calderon, O. H. (1967). *Dev. Ind. Microbiol.* **8**, 321–326.
Starkey, R. L., and Wight, K. M. (1945). "Anaerobic Corrosion of Iron in Soil." Amer. Gas Assoc., New York.
Stephenson, M., and Stickland, L. H. (1931). *Biochem. J.* **25**, 215–220.
Stranger-Johannessen, M. (1986). In "Biodeterioration 6" (S. Barry and D. R. Houghton, eds.), pp. 218–223. C.A.B. International Mycological Institute, Slough, England.
Stratfull, F. F. (1961). *Corrosion* **17**, 493–496T.
Syrett, B. C., MacDonald, D. D., and Wing, S. S. (1979). *Corrosion* **35**, 409–422.
Tatnall, R. E. (1981a). *Mater. Perform.* **20**, 41–48.
Tatnall, R. E. (1981b). *Mater. Perform.* **20**, 32–38.
Taylor, J., and Parkes, R. J. (1983). *J. Gen. Microbiol.* **19**, 3303–3309.
Thorpe, P. H. (1976). *Corros. Australas* **4**, 12–14.
Tiller, A. K. (1982). In "Corrosion Processes" (R. N. Parkins, ed.), pp. 115–159. Applied Science Publ., New York.
Tiller, A. K. (1983a). In "Microbial Corrosion," pp. 104–107. Metals Society, London.
Tiller, A. K., and Booth, H. (1968). *Corros. Sci.* **8**, 549–555.
Tuovinen, O. H., Button, K. S., Vuorinen, A., Carlson, L., Mair, D. J., and Yut, L. A. (1980). *J. Am. Water Works Assoc.* **72**, 626–635.
Tuttle, R. N., and Kane, R. D., (eds.) (1981). "H_2S Corrosion in Oil and Gas Production—A Compilation of Classic Papers." Nat. Assoc. Corros. Eng., Houston, Texas.

Uhlig, H. H. (1963). "Corrosion and Corrosion Control." Wiley, New York.
Umbreit, W. (1976). *Dev. Ind. Microbiol.* **17**, 265–268.
Videla, H. A. (1981). "Corrosão Microbiologica." Biotechnologia, Vol. 4. Blücher, São Paulo.
Videla, H. A. (1986). *Int. Conf. Biol. Induced Corros.* pp. 215–222.
Videla, H. A., and Salvarezza, R. C. (1984). "Introduccion a la Corrosion Microbiologica." Mosaico Libreria Agropecuaria, Buenos Aires.
Wakerley, D. S. (1979). *Chem. Ind.* **19**, 657–659.
Walch, M., and Mitchell, R. (1986). *Int. Conf. Biol. Induced Corros.* pp. 201–208.
White, D. C. (1983). *Symp. Soc. Gen. Microbiol.* **34**, 37–66.
White, D. C., Nivens, D. E., Nichols, P. D., Mikell, A. T., Jr., Kerger, B. D., Henson, J. M., Geesey, G. G., and Clarke, C. K. (1986). *Int. Conf. Biol. Induced Corros.* pp. 233–243.
Whitney, W. R. (1903). *J. Am. Chem. Soc.* **22**, 394–406.
Widdel, F., and Pfennig, N. (1977). *Arch. Microbiol.* **112**, 119–122.
Wilkinson, T. G. (1983). In "Microbial Corrosion," pp. 117–122. Metals Society, London.
Willingham, C. A., and Quinby, H. L. (1971). *Dev. Ind. Microbiol.* **12**, 278–284.
Worthingham, R. G., Jack, T. R., and Ward, V. (1986). *Int. Conf. Biol. Induced Corros.* pp. 339–350.

Economics of the Bioconversion of Biomass to Methane and Other Vendable Products

RUDY J. WODZINSKI,* ROBERT N. GENNARO,*
AND MICHAEL H. SCHOLLA[†]

*Department of Biological Sciences
University of Central Florida
Orlando, Florida 32816

[†]Department of Biological Sciences
Memphis State University
Memphis, Tennessee 38152

I. Introduction

When the energy shortage of the 1970s occurred, it kindled an unprecedented interest in anaerobic digestion. Basic and applied studies were undertaken on processes to bioconvert various types of energy crops or waste substrates to methane. Many of these studies had a secondary goal of utilizing the value of the methane produced to abate pollution in an economical manner.

Earlier studies on the economics of these processes indicated that the value of the methane alone did not usually yield a profit incentive sufficient to entice many investors to underwrite the construction of methane-producing anaerobic digestion facilities. It became obvious that if profitable anaerobic digestion to produce methane was to be implemented in the near future, the process would require additional vendable products, subsidies or credits for abating pollution, or a combination of all possibilities.

A number of possibilities for vendable products and credits from anaerobic digestion have been suggested. Economic studies have been performed on the possibility of compressing the biogas and separating the carbon dioxide for sale (Wodzinski and Gennaro, 1979; Hinley, 1979), refeed of digester protein from dairy and steer fermentation (Varel et al., 1977; Ashare et al., 1977), direct sale of digester solids as fertilizer (Ashare et al., 1977), upgrading digester residues with supplemental nitrogen, phosphorus, and potassium to produce "organic fertilizer" (Wodzinski and Gennaro, 1979; Sartain, 1979), and on-site use in marketing of electricity from digester-produced methane (Rodriguez, 1979). It is also obvious that at this time the economics of the system do not support the use of substrate which is expressly produced for anaerobic digestion.

Since the onset of the renewed interest in anaerobic digestion in the 1970s, many of the key economic factors that affect the profitability of methane production have changed. It is, therefore, timely to reevaluate the economics of the production of methane and other vendable products by anaerobic digestion and to attempt to identify the areas of research that might enhance the wide-scale implementation of this process.

II. Input Data

A. GENERAL

The aim of this study was to subject the sensitive cost factors (both technical and economic) which influence the profitability of anaerobic digestion yielding methane and other vendable products to the economics of energy production operative in 1985. We have attempted, whenever possible, to use process conditions which give the highest yields of vendable products and the lowest obtainable capital costs (Tables I and II) for equipment which will perform over long periods of time with minimal maintenance costs. A further constraint on the input data that we used was that the data were derived from pilot-scale studies that had been verified as applicable to consistent performance of the process over relatively long periods of time.

III. Overview of Computer Programs

A Fortran program developed by Ashare et al. (1977) at Dynatech was modified extensively to reflect current data, credits for various potential by-products, and third-party financing (Brenner, 1983) that encompasses present (1985) tax law. The program has been revised a number of times since 1979 to encompass the advances made in rates of methane digestion and in hardware. The program as it is currently constituted retains the format established by Ashare et al. (1977) and calculates reactor performance; mass balances for liquids, solids, and gases; energy balances and needs; capital costs; and manufacturing costs.

Essentially the modified program calculates the after-tax benefits (ATB) and the costs of methane production which would accrue if an anaerobic digestion system were implemented to digest dairy, beef, swine, or poultry wastes. It has been further modified to calculate the after-tax benefits and methane value if water hyacinths (*Europa crassipes*) are digested. The program allows for various scenarios of credits to be taken, such as for carbon dioxide, fertilizer, feed, wastewater treatment, cogeneration of electricity, and use of "waste heat." The Hooke–Jeeves pattern-move portion of the Ashare et al.

TABLE I

INPUT BASELINE VARIABLES COMMON TO DAIRY, BEEF FEEDLOT, SWINE, POULTRY, AND HYACINTH PROGRAMS

Number of operators	2.5
Salary ($/hour)	7.5
Water cost ($/$10^3$ gallons)	0.40
Steam cost ($/$10^6$ BTU)	4.5
Electricity cost ($/kW-hr)	0.0752
Digester cost factor	0.0636
Vacuum filter cost factor	0.3206
Gas compressor cost factor	0.1955
Heat exchanger cost factor	3.07
Instrument cost factor	41.28
Generator cost factor	7.64
Site preparation cost factor	24.25
Maintenance–supply factor	758.5
Fraction interest debt	0.10
Fuel escalation rate	0.10
Inflation rate	0.05
Tax rate	0.46
Life of project (years)	10
Federal credits	0.20
Cost of urea ($/g)	5.27×10^{-4}
Cost of isobutylenediurea ($/g)	1.29×10^{-3}
Cost of P_2O_5 ($/g)	9.0×10^{-5}
Cost of K_2O ($/g)	9.4×10^{-5}
Desired N in fertilizer (g/kg)	60
Desired P in fertilizer (g/kg)	60
Desired K in fertilizer (g/kg)	60
Wholesale price of CO_2 ($/ton)	11
Gram fraction CO_2 (48–60°C)	0.043
Gram fraction CO_2 (25–42°C)	0.30
Gram fraction CH_4 (48–60°C)	0.57
Gram fraction CH_4 (25–42°C)	0.70
Electricity value to power grid ($/kW-hr)	0.0366
Number of fertilizer plant operators	1
Number of fertilizer plant chemists	1
Salary of fertilizer plant operator ($/hour)	7.5
Salary of fertilizer plant chemist ($/hour)	12.5
Insurance for fertilizer plant ($)	16,547.00
Digester plastic bag and pump cost factor	24.58
Pollution factor for nitrogen removal (tons/year)	8.66×10^{-3}
Pollution factor for phosphorus removal (tons/year)	2.96×10^{-3}
Wholesale value of CH_4 ($/$10^3$ ft^3)	3.00

TABLE II

INPUT BASELINE VARIABLES

Variable	Dairy	Beef	Swine	Poultry	Hyacinths
Raw materials cost factor	31.38	26.46	1×10^{-4}	0	31.38
NH_3–N in effluent (mM)	166	166	210	166	166
NH_3–N in liquid effluent (mM)	86	86	105	86	86
P in effluent (mM)	25.7	25.7	0.1	25.7	25.7
P in liquid effluent (mM)	0	0	0	0	0
K in effluent (mM)	52	0	0.1	0	52
K in liquid effluent (mM)	0	0	0	0	0
Fraction N in digested solids	0.066	0.066	0.08	0.066	0.066
Fraction P in digested solids	0.0257	0.0257	0.01	0.03	0.0257
Fraction K in digested solids	0.0563	0.0563	0.01	0.03	0.0563
Substrate cost ($/g)	2.76×10^{-6}	2.76×10^{-6}	2.76×10^{-6}	2.76×10^{-6}	0

(1977) program was modified to allow for the calculation of least-cost economics via a brute force iteration technique. An overview of the program is in Table III.

This technique was used to perform cost sensitivity analyses and determines the effects of various process parameters if the reaction rate constant is increased and if the refractory volatile solids of dairy wastes are varied. The purpose of these particular studies was to ascertain the potential value of research to increase the rate of methane production and/or the breakdown of refractory solids in dairy wastes.

A reproduction of one of the computer printouts of the program used for water hyacinths is in Table IV. In this example a conventional digester was used and a centrifuge was used to concentrate solids. None of the heat from the effluent was recovered and the gas was used internally. The printout illustrates the data obtained on a 10-acre water hyacinth pond. It includes the feed makeup process conditions, heat balance, power requirements, capital costs, manufacturing costs, including credits for fertilizer and for

TABLE III

OVERVIEW OF LOGIC FLOW FOR COMPUTER PROGRAM

I. Read in constants and options
II. Determine reaction rate constants
III. Determine reactor performance (brute force iteration) for plug flow reactor or continuous stirred reactor
IV. Determine gas production
V. Determine material mass balance
    A. Flow rate
    B. Total solids
    C. Substrate
    D. Water flow rate
    E. Nonvolatile solids
    F. Outlet total solids
        1. Centrifuge (mass of centrifuge solids)
           or
        2. No centrifuge
    G. Bleed stream
    H. Recycle
    I. Makeup water
    J. Digester volume
    K. Annual gas production
VI. Calculation of heat exchange area
VII. Determine energy balances
    A. Heat loss of digester
    B. Heat of reaction
    C. Heat loss due to water evaporation
    D. Heat loss due to bleed stream (liquid and solids)
        1. With heat exchanger
        2. Without heat exchanger
    E. Heat loss with gas
    F. Heat loss due to water makeup
    G. Heat to operate fertilizer plant
        1. With fertilizer plant: Excess heat from electrical generator to dry fertilizer
        2. No fertilizer plant
    H. Heat loss from slurry
        1. With heat exchanger
        2. Without heat exchanger
    I. Substrate heat loss
    J. Determine horsepower requirements
        1. Horsepower requirement for mixer or slurry
        2. Horsepower requirement for digester if mixed
        3. Horsepower requirement for centrifuge
        4. Horsepower subtotal
    K. Determine on-site energy uses (option)
        1. Annual amount of gas used to produce electricty
        2. Annual amount of gas used to heat water
        3. Total internal annual gas usage
        4. Excess heat produced by on-site generation of electricity

(continued)

TABLE III *(Continued)*

- L. Calculation of total heat needs
- M. Calculation of horsepower required to compress gas
- N. Calculation of total horsepower requirements
- VIII. Program cost
  - A. Read in variables
  - B. Determine capital cost of components
    1. Digester cost
       a. Conventional, includes digester heat exchanger, insulation, and valves
       b. Plastic, includes pumps
    2. Centrifuge cost if used
    3. Gas compressor, storage tanks, meters, pressure-relief valves, and switches
    4. Sludge drying bed cost (if used)
    5. Heat exchanger cost (effluent)
    6. Fertilizer plant cost
    7. Site preparation cost
    8. Harvester, grinder, and pumps (for hyacinths only)
  - C. Calculate plant investment cost
    1. Contractors fee cost
    2. Engineering fee cost
    3. Cost to confine cattle (if used)
    4. Gas compression piping (if used)
    5. Sewer pipeline for substrate transport (if used)
    6. Capital to construct ponds (hyacinths only)
    7. Calculate subtotal plant investment cost
    8. Project contingency cost
    9. Calculate total plant investment cost
  - D. Calculate annual manufacturing cost
    1. Raw material cost
    2. Determine if fertilizer plant is desired and if solids production is adequate
       a. Determine amount of NPK needed for desired levels if fertilizer used
       b. Determine cost of NPK needed for desired levels if fertilizer used
       c. Determine annual water cost
       d. Determine annual steam cost
       e. Determine annual utility cost
       f. Determine annual labor cost
       g. Determine annual administrative overhead costs
       h. Determine annual supplies cost
       i. Determine annual maintenance–supply cost
       j. Determine annual taxes and insurance costs
       k. Determine operating cost for fertilizer plant (if used)
       l. Determine fertilizer plant insurance cost
       m. Calculate annual gross operating cost
       n. Calculate credits for vendable products other than CH_4
          1. Fertilizer credit (if used)
          2. CO_2 credit (if used)
          3. Electricity credit (if used)

(continued)

ECONOMICS OF BIOMASS BIOCONVERSION

TABLE III (Continued)

        4. Calculation for pollution abatement (removal of N and P in hyacinth wastewater system)
     o. Calculate total net operating cost
E. Calculate total capital requirement
    1. Plant capital needed
    2. Determine federal tax credits
    3. Startup cost
    4. Working capital cost
    5. Total capital requirement (no federal credits)
    6. Total capital requirement (with federal credits)
    7. Annual capital cost with interest
    8. Total capital cost (present value over project life)
F. Calculate profitability
    1. Depreciation
    2. Operating cost over project life
    3. Total energy production present value
    4. Total energy production value over project life
    5. Rate of return
    6. Average cost of energy (present value)
    7. Payback period
    8. Gross revenue (yearly over project life and totals)
    9. Operating and maintenance (yearly over project life and totals)
   10. Debt service (yearly over project life and totals)
   11. Cash flow (yearly over project life and totals)
   12. Interest (yearly over project life and totals)
   13. Depreciation (yearly over project life and totals)
   14. Profit/loss (yearly over project life and totals)
   15. Tax credit (yearly over project life and totals)
   16. Total after-tax benefits (yearly over project life and totals)
G. Calculate potential pollution abatement

nitrogen and phosphorus removal from wastewater, and economic analyses of the digester system, annual methane production, value of the methane, and digester operating costs. It lists the revenues, costs, debt service, cash flow, interest, depreciation, credit or loss, federal credits, and after-tax benefits. These are listed for each year over the life of the project, and for an additional year after the system has been fully depreciated and the entire cost of the plant is paid. Also, 10-year totals (the life of the project) for each of the economic parameters are calculated. In the particular example illustrated, a profit of 3.43×10^6 was realized over the 10-year period with an internal rate of return of 306.8%. Most of the profit (ca. 3×10^6) was derived from the production of fertilizer and wastewater treatment because the value of the methane produced was only $385,000.

TABLE IV

Sample Calculation Using Modified Computer Program: Optimum Design for a Continuous-Stirred Tank Digester[a]

CONVENTIONAL DIGESTER DESIGN
WITH A CENTRIFUGE
WITHOUT A HEAT EXCHANGER
GAS USED INTERNALLY ONLY

FEED MAKEUP

NUMBER OF ACRES=	10.	
RATE OF PLANT PRODUCTION=	246.70	LBS OF SOLIDS/DAY/ACRE
MOISTURE CONTENT OF PLANTS=	90.7	PERCENT
VOLATILE SOLIDS=	85.0	PER CENT OF TOTAL SOLIDS
REFRACTORY VOLATILE SOLIDS=	33.0	PER CENT OF VOLATILE SOLIDS

PROCESS CONDITIONS AT MAXIMUM RATE OF RETURN

FLOW STREAMS

BLEED STREAM=	0.24330E+04	GALLONS/DAY
RECYCLE STREAM=	0.73113E+04	GALLONS/DAY
WATER MAKE-UP=	0.00000E+00	GALLONS/DAY
TEMPERATURE OF MAKE UP WATER=	20.	DEG. C
CENTRIFUGE PELLET=	0.31212E+01	TONS /DAY
SOLIDS IN CENTRIFUGE PELLET=	25.0	PER CENT
VOLATILE SOLIDS CONVERSION	39.3	PER CENT

HEAT BALANCE

TEMPERATURE OF SURROUNDINGS=	20.	DEG. C
HEAT LOSS FROM DIGESTER=	0.10895E+07	BTU/DAY
EVAPORATIVE COOLING=	0.43149E+05	
HEAT LOSS WITH BLEED=	0.54839E+06	
HEAT LOSS WITH GAS	0.94895E+04	
HEAT LOSS WITH CENT. SOLIDS=	0.16877E+06	
HEAT IN WITH MAKEUP WATER=	0.00000E+00	
HEAT OF REACTION=	0.53927E+06	
HEAT LOSS WITH SLURRY	0.00000E+00	
WASTE HEAT FROM PLANT GEN=	0.38477E+06	
WASTE HEAT FROM FARM GEN=	0.16049E+07	
HEAT TO DRY FERTILIZER=	0.12784E+07	
TOTAL HEAT REQUIREMENTS=	0.60875E+06	BTU/DAY

POWER REQUIREMENTS

DIGESTOR, MIXING =	0.27786E+00	
GAS COMPRESSOR=	0.00000E+00	
CENTRIFUGE=	0.41543E+00	
TOTAL POWER REQUIREMENTS=	0.69329E+00	HP

(continued)

TABLE IV (Continued)

ECONOMIC ANALYSIS OF THE DIGESTER SYSTEM

CALCULATION OF TOTAL PLANT INVESTMENT

CAPITAL INVESTMENT (IN DOLLARS)

DIGESTER(1.0 NEEDED),INCLUDES DIGESTER HEAT EXCHANGER, INSULATION AND VALVES
```
                                         0.12311E+06 PER UNIT
HEAT EXCHANGER EFFLUENT                  0.00000E+00
CENTRIFUGE                               0.48372E+05
GAS COMPRESSOR, STORAGE TANKS, METERS,
    PRESSURE RELIEF VALVES AND
    SWITCHES                             0.00000E+00
SLUDGE DRYING BED                        0.00000E+00
INSTRUMENTATION CONTROLS AND ELECTRICAL  0.34442E+05
ELECTRICAL GENERATOR                     0.44486E+03
FERTILIZER PLANT                         0.17999E+06
SITE PREPARATION                         0.34603E+05
HARVESTER,GRINDER,PUMPS                  0.18394E+05

TOTAL CAPITAL INVESTMENT                 0.43936E+06
CONTRACTORS FEE(=0.1*TOT. CAP. INV.)     0.43936E+05
ENGINEERING(=0.05*TOT.CAP.INV.)          0.21968E+05

GAS COMPRESSION PIPING                   0.00000E+00
SEWER FOR MANURE TRANSPORT               0.00000E+00
CAPITAL TO CONSTRUCT PONDS               0.25000E+06

SUBTOT. PLANT INVESTMENT                 0.75526E+06
PROJECT CONTINGENCY(=0.15*SUB. PL. INV.) 0.11329E+06

TOTAL PLANT INVESTMENT                   0.86855E+06
```

MANUFACTURING COST(IN DOLLARS/YEAR)

```
  COST TO SUPPLEMENT SOLIDS W/NPK        0.88990E+05

  WATER MAKEUP(AT 0.40 DOL./M GAL)       0.00000E+00
  STEAM(AT 4.50DOL./MM BTU)              0.00000E+00
  POWER(AT0.075DOL./KWH)                 0.00000E+00

  OPERATING LABOR( 3.0 MEN AT 7.50 DOL./HR)  0.51900E+05
  MAINTENENCE LABOR(=0.015*TOT. PL. INV.)    0.13028E+05
  ADMIN. OVERHEAD(=0.6*OPER.+MAINT. LABOR)   0.38957E+05
  OPERATING SUPPLIES(=0.3*OPER. LABOR)       0.15570E+05
  MAINTENANCE SUPPLIES(=0.015*TOT. PL. INV.) 0.13028E+05
  LOCAL TAXES+INSURANCE(=0.027*TOT. PL. INV.) 0.23451E+05
  TOTAL FERTILIZER PLANT LABOR
      1.OPERATORS AT 7.50 DOL./HR AND
      1.CHEMISTS AT 12.50 DOL/HR
          TOTAL EQUALS                   0.55360E+05
  FERTILIZER PLANT INSURANCE             0.16547E+05
  ANNUAL, HARVEST LABOR,GRINDING,FUEL COST  0.22512E+05

  TOTAL GROSS OPERATING COST             0.33934E+06
  CREDIT FOR FERTILIZER                  0.74966E+06

  CREDIT FOR CO2                         0.00000E+00

  CREDIT FOR FARM ELECTRICITY            0.00000E+00

  CREDIT FOR N AND P REMOVAL             0.27430E+05

  TOTAL NET OPERATING COST              -0.43774E+06
```

WHEN TOTAL NET OPERATING COST IS NEGATIVE,THE CREDITS
FOR CO2 AND FERTILIZER EXCEED THE TOTAL NET OPERATING COST

(continued)

TABLE IV (Continued)

ECONOMIC ANALYSIS OF DIGESTER SYSTEM

TOTAL PLANT INVESTMENT (DOLLARS)	0.86855E+06
FEDERAL CREDITS	0.19076E+06
START-UP(0.2*TOT. GR. OPER. COST)	0.67869E+05
WORKING CAPITAL(0.02*TOT. PL. INV.)	0.17371E+05
TOTAL CAPITAL REQUIREMENT(NO FED. CREDITS)	0.95379E+06
TOTAL CAPITAL REQUIREMENT(WITH FED. CREDITS)	0.76303E+06
ANNUAL CAPITAL COST WITH INTEREST	0.11176E+06
TOTAL CAPITAL COST(PRESENT VALUE OVER PROJECT LIFE)	0.88440E+06
DEPRECIATION(STRAIGHT LINE 10 YRS)	0.24365E+06
OPERATING COST OVER PROJECT LIFE	-0.43774E+07
TOTAL ENERGY ANNUAL VALUE	0.38562E+05
TOTAL ENERGY PRODUCTION OVER PROJECT LIFE	0.49100E+06
RATE OF RETURN	0.11918E+01
AVERAGE COST OF ENERGY($/M CU FT)	-0.29071E+03
PAY BACK PERIOD(YEARS)	0.30853E+02
ANNUAL GAS PRODUCTION	0.28748E+04 M CU. FT./YEAR
NET ANNUAL GAS PRODUCTION	0.12854E+04 M CU.FT./YEAR
NET ANNUAL CH4 PRODUCTION(@$3.00 M CU FT)	0.38562E+04 DOLLARS

DIGESTER OPERATING CONDITIONS

TEMPERATURE=	35.0	DEG. C
RETENTION TIME=	12.0	DAYS
VOLATILE SOLIDS TO DIGESTER=	1.49	LBS/CU FT.
PERCENT SOLIDS TO DIGESTER=	2.8	
NUMBER OF DIGESTERS=	1.	
VOLUME PER DIGESTER=	0.16841E+02	
GAS PRODUCTION=	0.28748E+04	M CU.FT./YEAR

IMPACT ON RELIEVING POLLUTION FROM SEC. EFFLUENT

NITROGEN REMOVED=	0.19162E+04	TONS/YEAR
PHOSPHORUS REMOVED=	0.45625E+03	TONS/YEAR

(continued)

TABLE IV (Continued)

YEAR	GROSS REVENUES	O&M COSTS	DEBT SERVICE	CASH FLOW	INTEREST	DEPRE-CIATION	PROFIT LOSS	TAX CREDIT	TOTAL ATB
1	0.7809E+06	0.3393E+06	0.1118E+06	0.3298E+06	0.6867E+05	0.1004E+06	-.27252E+06	0.19076E+06	0.37343E+06
2	0.8590E+06	0.3563E+06	0.1118E+06	0.3910E+06	0.6436E+05	0.1473E+06	-.29110E+06	0.00000E+00	0.23777E+06
3	0.9449E+06	0.3741E+06	0.1118E+06	0.4591E+06	0.5962E+05	0.1406E+06	-.37063E+06	0.00000E+00	0.25892E+06
4	0.1039E+07	0.3928E+06	0.1118E+06	0.5348E+06	0.5441E+05	0.1406E+06	-.45163E+06	0.00000E+00	0.29096E+06
5	0.1143E+07	0.4125E+06	0.1118E+06	0.6191E+06	0.4868E+05	0.1406E+06	-.54166E+06	0.00000E+00	0.32664E+06
6	0.1258E+07	0.4331E+06	0.1118E+06	0.7129E+06	0.4237E+05	0.0000E+00	-.78225E+06	0.00000E+00	0.29044E+06
7	0.1383E+07	0.4548E+06	0.1118E+06	0.8170E+06	0.3543E+05	0.0000E+00	-.89331E+06	0.00000E+00	0.33459E+06
8	0.1522E+07	0.4775E+06	0.1118E+06	0.9326E+06	0.2779E+05	0.0000E+00	-.10165E+07	0.00000E+00	0.38364E+06
9	0.1674E+07	0.5014E+06	0.1118E+06	0.1061E+07	0.1940E+05	0.0000E+00	-.11533E+07	0.00000E+00	0.43813E+06
10	0.1841E+07	0.5264E+06	0.1118E+06	0.1203E+07	0.1016E+05	0.0000E+00	-.13048E+07	0.00000E+00	0.49862E+06
11	0.2026E+07	0.5528E+06	0.0000E+00	0.1473E+07	0.0000E+00	0.0000E+00	-.14728E+07	0.00000E+00	0.67749E+06

10 YR TOTALS

GROSS REVENUES	O&M COSTS	DEBT SERVICE	CASH FLOW	INTEREST	DEPRE-CIATION	PROFIT LOSS	TAX CREDIT	TOTAL ATB
0.1245E+08	0.4268E+07	0.1118E+07	0.7060E+07	0.4309E+06	0.6694E+06	-.70777E+07	0.19076E+06	0.34291E+07

INTERNAL RATE OF RETURN % = 306.82

^{a}M = 1000; MM = 1,000,000.

The following summary of the results has been developed from over 1000 scenarios which have been run using computer outputs generated with the modified program.

IV. Substrates for Anaerobic Digestion

Although many substrates can be bioconverted to methane, only a few are considered in potential digester systems. Methane has a low value as compared to other fermentation products. This low value imposes severe economic constraints on any potential process. Substrates must be fermentable and available at a no or minimal cost throughout the year, in large quantities and in relatively concentrated form. Animal wastes meet these criteria. Water hyacinths that are used to treat wastewater are also a likely substrate. Energy crops, i.e., plant materials that are cultivated for the express purpose of serving substrates to produce methane, are too expensive when production and harvest costs are added to the cost of producing methane.

Overcash et al. (1983) have listed the quantity, availability, and locations of various livestock wastes in the United States. The total potential amount of biogas that would be produced if all of these substrates were subjected to anaerobic digestion would be 2.63×10^9 ft^3 (Table V).

It is necessary to assume some value for the wastes that are produced. Usually $2.50 per wet ton of dairy manure is considered a minimal fertilizer value. This value is used in the economic calculations for all of the livestock wastes and includes transfer of the wastes to the digester. The substrate cost using the programs with hyacinths was zero since hyacinth harvesting and processing (chopping) costs are relatively high and were calculated separately.

TABLE V

The Number of Animals Produced per Year and Potential Biogas Production[a]

Animals produced/year	Number of animals in the United States	Potential biogas production (ft^3/head/day)	Potential total biogas production (ft^3/head/day)
Dairy milk cows (1976)	1.1×10^7	46.4	5.1×10^8
Beef cattle (1973)	3.1×10^7	27.6	8.56×10^8
Hogs and pigs (1973)	8.33×10^7	3.9	3.25×10^8
Broilers (1976)	3.25×10^9	0.29	9.4×10^8

[a]Source: Overcash et al. (1983).

V. Capital Costs

Various companies throughout the United States were contacted for capital costs for the various types of equipment needed to bioconvert the various substrates to methane and other vendable products. These cost data were incorporated into the revised computer programs. The Marshall–Stevens cost index was utilized to update costs whenever necessary to reflect 1985 values.

VI. Process Conditions

Equations from the Ashare et al. (1977) program were used to determine the material balances and the kinetics of methane production. However, the process parameters that were used to calculate the economics for the various substrates were derived from various pilot studies that were conducted on a meaningful scale for a long enough period to establish steady-state conditions. The rate and total amount of methane produced from the various substrates as calculated by the computer program were made to agree with those actually observed by varying elements in the equations, such as the rate constant and the ratio of grams of chemical oxygen demand (COD) per grams biodegradable volatile solvents. The process parameters that were used are in Table VI.

VII. Digester, Gas, Solids, and Liquid Effluent Process Designs

Several alternative processes to digest substrates to methane and other vendable products might be practical. A schematic design for an integrated methane limit which includes several alternatives is shown in Fig. 1. A modular digester system based on the data of Coppinger et al. (1978), and later refined by her (unpublished) for Anaerobic Energy Systems Inc., was utilized for capital data and process conditions because it has been implemented for the digestion of dairy manure to methane at a biogas research facility at the Otter Run Farm, Virginia. It consists of site preparation, an underground settling basin, a centrifugal chopper pump, epoxy-coated steel tanks (with grit removers) bolted together with stainless-steel bolts, an insulated external heat exchanger panel on the tank walls (for heating), pressure-relief valves, a vacuum breaker, insulated pipelines, a gas handling system that filters, meters, and compresses the gas to 240 psi, propane storage tanks, water removal traps, hydrogen sulfide scrubbers and filters, control panels, and safety equipment. The digestion system and the costs associated with it were used for all computer scenarios except those which employed plastic bag digesters.

TABLE VI

Process Parameters Used to Calculate Heat, Solids, Liquids, and Gas Balances[a]

Process parameter	Dairy wastes[b]	Feedlot wastes[c]	Swine wastes[d]	Poultry wastes[e]	Water hyacinths[f]
Total solids produced (g/day/animal or g/day/acre)	4800	4600	5113	31.0	112,000
Fractional volatile solids	0.82	0.90	0.83	0.76	0.85
Fractional water content	0.875	0.875	0.97	0.80	0.907
Fractional refractory volatile solids	0.524	0.524	0.38	0.33	0.33
Ratio of grams COD/grams biodegradable volatile solids	1.57	1.31	2.23	1.52	1.79
Temperature of digestion (°C)	35	35	35	35	35
Concentration influent total solids (%)	10	10	4.87	3.94	2.82
Rate constant	5.92×10^9	5.92×10^9	3.85×10^{12}	1.02×10^{11}	5.92×10^9

[a]The following parameters were used for all substrates: ambient temperature, 20°C; Arrhenius activation energy, 7588/cal/mol; fractional volatile solids in filter cake, 0.25.
[b]Source: Wodzinski et al. (1986).
[c]Source: Hashimoto (1982); Varel et al. (1980).
[d]Source: Fisher et al. (1979).
[e]Source: Steinberger and Shih (1984).
[f]Source: Chynoweth et al. (1982).

Cases in which enough carbon dioxide would be produced (25 tons/day), equivalent to the digestion of the manure from 16,943 head of dairy cattle, warranted separation of the methane by compression to 2,000 psi; these cases were used for the cost input and the energy requirements to run the compressor. In these instances, the methane was pipeline quality and could be used either on site or run into a pipeline. Methane or biogas could also be used on site in generators; the resultant electricity furnished to the power grid and the waste heat, 185°F water with a recovery efficiency of 90%, used on site. The capital costs and specifications associated with the digester system were provided by Anaerobic Energy Systems Inc. in 1981 and were inflated to 1985 costs by use of the Marshall–Stevens Index.

A. Fertilizer Plant Design for Use of Digester Solids

Design data and sizing of a potential 5-ton per day modular fertilizer plant (Fig. 2) were provided by Edmon A. Glos II of Bonne Terre Ltd., Winter Park, Florida. In the proposed process, the digester effluent is centrifuged to collect solids and the solids are partially dried in a sludge

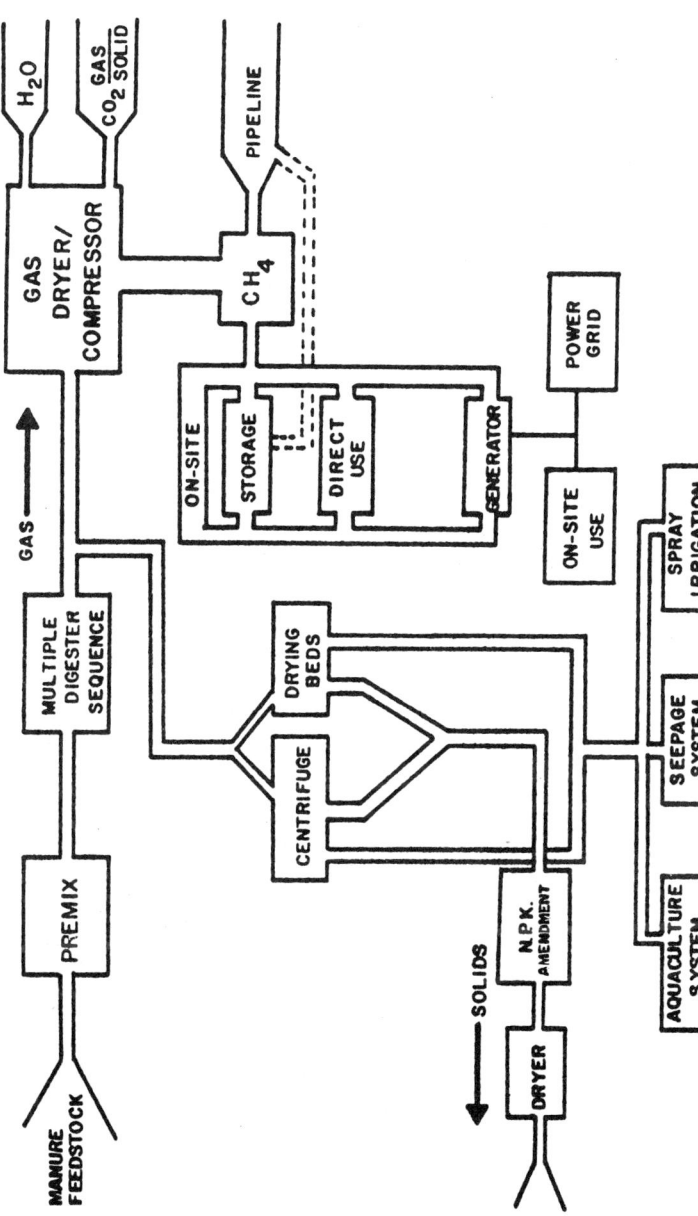

FIG. 1. Schematic diagram of options available for digestion of substrate to methane and other vendable products.

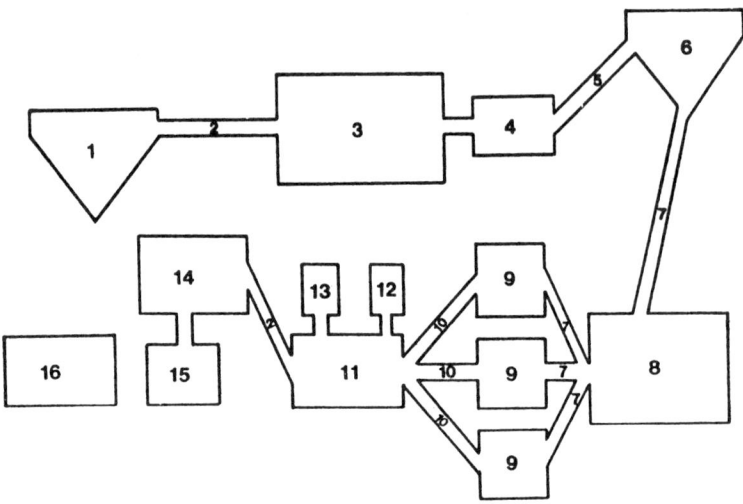

FIG. 2. Schematic diagram of a fertilizer plant designed to convert digester solids to 6-6-6 organic fertilizer. (1) Dry sludge feed hopper, (2) cold feed conveyor, (3) kiln (dryer), (4) vibrator screen (¼′ mesh), (5) bucket elevator, (6) hot hopper storage, (7) hot conveyor, (8) jacketed Bock mill (2-ton capacity), (9) hot hopper storage, (10) speed belt feeder (24″ wide), (11) pug mill cold feed (2000-ton capacity), (12) sulfuric acid (20–40 gallons/minute, 1000-gallon fiberglass tank), (13) urea formaldehyde of isobutylenediurea reservoir with 20′ head (18,000-gallon stainless-steel tank), (14) storage building (40′ × 60′ × 20′ high, open sides, with drape-cloth sides), (15) automatic bagger, and (16) storage.

drying bed; they then enter a feed hopper and are conveyed to a kiln dryer, sized on a vibrator screen, elevated into a feed hopper, and run through a 2-ton-capacity jacketed Bock mill. The solids are placed into a hot hopper and are conveyed to a pug mill. Either urea, urea formaldehyde, or isobutylenediurea is added as a source of nitrogen. Phosphorus and potassium are added as needed. After leaving the pug mill, the material is bagged and stored. Waste heat produced during electricity generation from biogas was utilized whenever possible for drying the solids.

The capital cost for the fertilizer plant and the material balances were provided by Mr. Glos. The costs of fertilizer amendments were direct price quotations from large fertilizer producers.

B. Water Hyacinth Production in Water Effluents

Input cost data for the production of water hyacinths in ponds utilizing secondary wastewater and values for credits for the removal of nitrogen and phosphorus from the wastewater were provided by Amasek, Inc., Cocoa Beach, Florida, and were based on data obtained from an ongoing pilot operation at the Iron Bridge Wastewater Treatment Plant, Orlando, Florida.

VIII. Factors Affecting Profitability of Dairy Cattle Waste Digestion

A. THE EFFECT OF HERD SIZE

As one would expect, economics of scale are realized in the digestion of dairy manure to vendable products. In the "best case" scenario, that is, production and sale of methane at $3 per thousand cubic feet (Mft^3), production of 6-6-6 organic fertilizer with a wholesale value of $140/ton, credit for on-farm use of the electricity, utilization of waste heat generated by electrical production, utilization of federal investment tax credits and energy tax credits, and accelerated depreciation, the break-even point for a profitable digester operation is approximately a herd size of 948 animals (Table VII). The profitability increases very rapidly as the herd size is increased. These data indicate that under present economic conditions digestion system are not warranted in most family dairy operations because the herd sizes are inadequate for profitability. These results are contrary to those reported earlier under different economic conditions (Jewell et al., 1976, 1978).

B. THE EFFECT OF FERTILIZER CREDITS

A comparison of the average cost of methane (ACM) production among cases in which fertilizer is used and no credit is taken (Table VII) indicates that in every case it is not profitable to run a methane digester if the solids are not recovered. All credits for vendable products other than methane are formulated into a program to reduce the cost of methane production. Since the present wholesale value is $3/Mft^3, a value for the ACM over $3/Mft^3 is not profitable. An ACM below $3/Mft^3 is certainly profitable—the more negative the value, the greater the profitability. It should be noted that the ACM does not include federal tax benefits nor the accelerated cost recovery (ACR) method of depreciation. If an installer has third-party financing or has a very profitable operation that includes a huge tax liability, then the federal tax benefits become very meaningful; for example, an ATB of $5900 is realized with a herd size of 1000 even though the ACM is calculated at $13.31/Mft^3 (Table VII).

The calculated net value of the fertilizer is at least 10 times as much as

TABLE VII

Effect of Number of Animals on Capital Requirements and Profitability of Dairy Waste Conversion to Vendable Products[a]

Herd size	Total capital[b] investment ($1000)	Total plant investment ($1000)	Net CH_4[c] produced/year ($1000)	Net fertilizer[d] produced/year ($1000)	Average[e] after-tax benefits ($1000)
100	200.2	207.8	1.9	2.3	−99.8
250	276.1	297.7	4.7	5.7	−83.3
500	383.9	424.7	9.5	115.2	−54.1
1000	578.9	653.8	19.1	230.5	5.9
2500	1102.1	1266.6	47.8	576.4	191.9
5000	1904.2	2203.2	95.6	1152.9	508.2
10,000	3423.8	3974.1	191.3	2305.7	1148.5
20,000	6330.7	7355.8	382.7	4611.5	2440.8

[a]Modified Ashere et al. (1977) computer program utilizing continuous-stirred tank digester; centrifuge; capital to confine cattle; credit for fertilizer; credit for electricity produced for on-farm use; 10% interest rate; 10% fuel escalation rate.
[b]With federal investment tax credits and energy tax credits.
[c]Methane value of $3/Mft^3.
[d]Organic 6-6-6 fertilizer value of $140/ton.
[e]Average of first 10 years.

the methane produced in a profitable operation (Table VII). Fertilizers are produced by recovering the solids from digestion and amending with organic nitrogen, phosphorus, and potassium; a slow release of the nitrogen produces a 6-6-6 fertilizer valued at $140/ton. It should be emphasized that the net value of the fertilizers has not been accomplished experimentally and should be tested before huge capital sums for plant installations are made.

C. The Effect of Cost to Confine Cattle on the Average Cost of Methane

Because confining cattle requires a capital investment of approximately $231 per head, it has a substantial effect on the profitability of any operation (Table VIII). Canton et al. (1979) has indicated that confining cattle and providing them shade from the sun in hot climates results in increased milk production and conception rates. If the benefits of increased milk production and conception are considered, and no cost for confining cattle is included in the capital investment of the digester facility, the digester operation becomes much more profitable. In a

TABLE VIII

EFFECT OF COST TO CONFINE CATTLE, SOLIDS HANDLING, AND FERTILIZER CREDITS ON THE AVERAGE COST OF METHANE (ACM) PRODUCTION FROM DAIRY WASTE[a]

Herd size	ACM ($/Mft³)	ACM ($/Mft³)	ACM ($/Mft³)	ACM ($/Mft³)
100	287.38	281.89	190.65	185.16
250	105.99	100.50	85.36	79.87
500	44.56	39.06	49.60	44.10
1000	13.31	7.82	31.37	25.87
2500	−6.06	−11.55	19.97	14.48
5000	−12.88	−18.37	15.91	10.42
10,000	−16.50	−22.00	13.72	8.22
20,000	−18.49	−23.98	12.49	7.03
Cost to confine cattle	Yes	No	Yes	No
Centrifuge	Yes	Yes	No	No
Fertilizer credit	Yes	Yes	No	No

[a]Modified Ashare et al. (1977) computer program utilizing a continuous-stirred tank digester; credit for electricity produced for on-farm use.

2500-head operation in which fertilizer credits are taken, the ACM is $11.55 when the cost to confine cattle is borne by the digester process and $6.06 when the cost to confine cattle is borne by the increased milk production and increased conception rates. If a dairy operation has its cattle confined or if even a small portion of the cost to confine cattle can be allocated to increased yields of milk and the remainder allocated to digester operations, it would result in increased profitability. Similar credits derived from confinement should be applicable in colder climates because energy to maintain body temperature should be less, especially if it is raining or snowing.

D. THE EFFECT OF CARBON DIOXIDE CREDITS

Early reports (Wodzinski et al., 1979; Hinley, 1979) indicated that there might be special situations when it is economically attractive to compress biogas to upgrade the methane produced and to produce carbon dioxide as an end product. Carbon dioxide producers will pay $4/ton of CO_2 and assume all the capital costs and recovery costs associated with carbon dioxide production. This value calculates to $0.49 per thousand cubic feet (Mft³) of carbon dioxide. By calculation, the methane has a value of about six times as much as the carbon dioxide.

Carbon dioxide producers place a lower limit of 25 tons per day carbon dioxide (Hinley, 1979) for economical recovery of the gas.

At 60°C, a temperature at which 57% methane and 43% carbon dioxide (Varel et al., 1977) are produced from dairy manure, it would take a herd size of 16,943 dairy cows to produce enough manure to achieve the 25 tons per day carbon dioxide in the digester process. An added benefit to compression of the gas to recover the CO^2 is that it would upgrade the biogas from 600 BTU/ft^3 to 1000 BTU/ft^3 and make it pipeline quality. This might be desirable where there is a natural gas transmission pipeline nearby. Gas transmission piping is expensive, $10,000 per inch-diameter per mile. Unless there is pipeline-quality methane produced in very large quantities, the cost of the pipeline feeders makes it prohibitive to feed into the main pipeline. In most instances, it is probably more economical to use the biogas produced by digestion directly to generate electricity on site and sell the electricity to the power companies, who feed it into their grids.

E. The Effect of Fuel Escalation Rate

Often in the calculation of the economics of energy production, factors are used to account for normal inflation rates for the overall economy and for fuel escalation rates. During the "energy crisis," the fuel escalation rate was much higher than the overall economy escalation rates. Accounting for these factors often enables one to predict when marginal energy production facilities might become profitable. This type of calculation is not entirely appropriate at the time of writing this review, because energy prices have been drastically reduced and the fuel escalation is lower than the overall inflation rate. In fact, it is a negative value. However, this type of calculation might be important in the future. We have used 5% as the overall inflation rate and compared the effective fuel escalation rate at 6, 8, and 10% (Table IX).

It is readily apparent that when the fuel escalation rate exceeds the normal inflation rate then the internal rate of return and the average after-tax benefits increase rapidly.

F. The Effect of Interest Rate

At the present time the prime rate for interest has been fluctuating around 10%. During the past 4 years, interest rates have fluctuated widely. When energy costs are high, interest rates are high. Because digestion processes include relatively high plant investment costs (Table VII), interest rates affect the profitability of methane production (Tables X and XI).

TABLE IX

THE EFFECT OF FUEL ESCALATION RATE ON INTERNAL RATE OF RETURN (IRR) AND AVERAGE AFTER-TAX BENEFITS (XATB) USING DAIRY WASTE[a]

	Fuel escalation rate					
	10%		8%		6%	
Herd size	IRR[b] (%)	$\overline{X}$ATB[c] ($1000)	IRR[b] (%)	$\overline{X}$ATB[c] ($1000)	IRR[b] (%)	$\overline{X}$ATB[c] ($1000)
100	−373.7	−97.9	−380.5	−99.6	−386.5	−101.2
250	−240.8	−78.5	−254.4	−82.9	−266.6	−86.9
500	−109.9	−44.5	−131.8	−53.4	−151.5	−61.4
1000	47.1	25.2	13.8	7.4	−16.0	−8.5
2500	288.8	240.0	235.4	195.6	187.2	155.6
5000	494.3	604.4	421.6	515.5	356.1	435.5
10,000	712.3	1340.7	617.8	1163.0	532.8	1003.1
20,000	939.4	2825.3	821.2	2469.8	714.8	2149.9

[a] Modified Ashare et al. (1977) computer program utilizing a continuous-stirred tank digester; centrifuge; no capital to confine cattle; credit for fertilizer; credit for electricity produced for on-farm use; 10% interest rate.

[b] IRR, Internal rate of return (total 10-year after-tax benefits − total 10-year debt service) × 100.

[c] $\overline{X}$ATB, Yearly average of total after-tax benefits over life of project.

IX. Factors Affecting Profitability of Beef Feedlot Waste Digestion

A. THE EFFECT OF HERD SIZE

If beef feedlot waste is digested anaerobically, the methane is sold at $3/Mft^3, the digester solids are recovered and amended with nitrogen, phosphorus, and potassium to produce a 6-6-6 "nonleachable, organic" fertilizer sold at $140/ton, electricity is generated and used on site, and the waste heat from electrical generation is used on site, then the break-even point for a profitable digester operation is a herd size of 948 animals (Table XII). The profitability increases rapidly as the herd size is increased. There are many feedlot operations in the United States that have herd sizes in excess of 1000 head (Overcash et al., 1983). The feedlot operators should analyze their specific situation to determine whether anaerobic digestion to methane and other vendable products would be profitable.

As would be expected, the profitability of beef feedlot waste digestion to methane and other vendable products closely parallels the profitability of dairy waste digestion. The two substrates are similar but

TABLE X

The Effect of Interest Rate on Internal Rate of Return (IRR) and Average After-Tax Benefits (XATB) Using Dairy Waste[a]

	Interest rate							
	8%		10%		12%		14%	
Herd size	IRR[b] (%)	X̄ATB[c] ($1000)	IRR[b] (%)	X̄ATB[c] ($1000)	IRR[b] (%)	X̄ATB[c] ($1000)	IRR[b] (%)	X̄ATB[c] ($1000)
100	−403.8	−96.8	−373.7	−97.9	−345.9	−98.9	−324.2	−100.0
250	−258.7	−77.3	−240.8	−78.5	−225.1	−79.8	−211.3	−82.2
500	−115.8	−43.0	−109.9	−44.5	−104.7	−46.2	−100.2	−47.9
1000	55.7	27.2	47.1	25.2	39.6	23.0	33.0	20.8
2500	319.6	243.2	288.8	240.0	261.9	236.6	238.2	233.2
5000	543.9	609.1	494.3	604.4	450.8	599.4	412.6	594.3
10,000	782.1	1348.1	712.3	1340.7	651.2	1333.1	597.7	1325.3
20,000	1030.1	2836.9	939.4	2825.3	860.1	2813.2	790.4	2800.7

[a] Modified Ashare et al. (1977) computer program utilizing a continuous-stirred tank digester; centrifuge; no capital to confine cattle; credit for fertilizer; credit for electricity produced for on-farm use; 10% fuel escalation rate.
[b] IRR, Internal rate of return (total 10-year after-tax benefits − total 10-year debt service) × 100.
[c] X̄ATB, Yearly average of total 10-year after-tax benefits.

TABLE XI

EFFECT OF INTEREST RATE ON THE AVERAGE COST OF METHANE
(ACM) PRODUCED FROM DAIRY WASTE[a]

	Interest rate			
	8%	10%	12%	14%
Herd size	ACM ($/Mft^3)	ACM ($/Mft^3)	ACM ($/Mft^3)	ACM ($/Mft^3)
100	279.15	281.89	284.70	287.67
250	99.13	100.50	101.91	103.38
500	38.21	39.06	39.94	40.85
1000	7.26	7.82	8.40	9.00
2500	−11.90	−11.55	−11.19	−10.82
5000	−18.63	−18.37	−18.11	−17.83
10,000	−22.20	−22.00	−21.80	−21.58
20,000	−24.14	−23.98	−23.82	−23.65

[a] Modified Ashare et al. (1977) computer program utilizing a continuous-stirred tank digester; no cost to confine cattle; with centrifuge; credit for electricity produced for on-farm use; credit for fertilizer.

there are differences in chemical composition, the amount of manure produced per animal per day, the fractional volatile solids, and the ratio of grams chemical oxygen demand per grams biodegradable volatile solids (Table VI). These factors influence the amount of methane produced from the manure per animal per day.

B. THE EFFECT OF FERTILIZER CREDITS, COST TO CONFINE CATTLE, CARBON DIOXIDE CREDITS, FUEL ESCALATION RATE, AND INTEREST RATE ON THE PROFITABILITY OF BEEF FEEDLOT WASTE CONVERSION TO VENDABLE PRODUCTS

In general, the same conclusions that were made for the effect of these parameters on profitability of anaerobic conversion of dairy wastes can be made for anaerobic conversion of feedlot wastes (Tables XII and XIII). The system financial performance on a yearly basis has been provided here for a 2500-head (Table XIV) and 10,000-head (Table XV) beef feedlot waste operation. These types of data have been generated

TABLE XII

EFFECT OF NUMBER OF ANIMALS ON CAPITAL REQUIREMENTS AND PROFITABILITY OF BEEF FEEDLOT WASTE CONVERSION TO VENDABLE PRODUCTS[a]

Herd size	Total capital[b] investment ($1000)	Total plant investment ($1000)	Net CH_4[c] production ($1000)	Net fertilizer[d] production ($1000)	Average[e] after-tax benefits ($1000)
250	174.1	220.1	4.4	48.3	−84.5
500	220.6	273.0	8.9	96.6	−56.7
1000	292.4	356.9	17.8	193.2	1.0
2500	455.6	550.3	44.5	483.1	179.1
5000	667.5	806.8	89.0	966.3	482.1
10,000	1013.0	1234.6	178.0	1932.6	1095.7
20,000	1588.6	1963.6	356.1	3865.3	2334.2
40,000	2564.4	3233.1	712.2	7730.0	4828.4
80,000	4255.9	5478.7	1424.5	15,461.0	9842.3
160,000	7251.9	9551.9	2849.0	30,922.0	19,908.0

[a] Modified Ashare et al. (1977) computer program utilizing a continuous-stirred tank digester; centrifuge; no capital to confine cattle; credit for fertilizer; credit for electricity produced for on-farm use; 10% interest rate; 10% fuel escalation rate.
[b] With federal investment tax credits and energy tax credits.
[c] Methane value of $3/Mft^3.
[d] Organic 6-6-6 fertilizer value of $140/ton.
[e] Average of first 10 years.

for each scenario for dairy, beef, swine, poultry, and water hyacinths. They clearly indicate on a year by year basis over the life of the project the cash flow, tax costs and credits, cost of money, depreciation, profit or loss, internal rate of return, and after-tax benefits. This information is invaluable as a basis for judging the economic feasibility of the anaerobic digestion process.

The after-tax benefits (ATB) that are realized in profitable operations are significant in the first year because they are directly related to energy and investment credits. Accelerated cost reduction (ACR) depreciation is taken from year 1 to year 5. Indeed, the ATBs usually decrease in the sixth year after the plant has been fully depreciated. In ensuing years, the ATBs increase due to the differential between the fuel escalation rate and the normal inflation rate. At this writing, these increases are not appropriate because the price of energy is decreasing whenever the "normal" inflation rate is being maintained. However, this is believed to be a temporary situation which will disappear after the "oil glut" is over.

TABLE XIII

EFFECT OF CONFINEMENT COST, SOLIDS HANDLING, FERTILIZER CREDITS, AND
INTEREST RATE ON AVERAGE COST OF METHANE (ACM)
PRODUCTION FROM BEEF FEEDLOT WASTE[a]

Herd size	ACM ($/Mft^3)	ACM ($/Mft^3)	ACM ($/Mft^3)	ACM ($/Mft^3)	ACM ($/Mft^3)	ACM ($/Mft^3)	ACM ($/Mft^3)
250	118.87	112.96	91.23	85.33	111.51	114.40	166.02
500	52.91	47.01	52.84	46.94	46.11	47.94	48.91
1000	19.34	13.43	33.23	27.33	12.84	14.04	14.67
2500	−1.45	−7.35	21.10	15.11	−7.71	−6.98	−6.59
5000	−8.75	−14.65	16.67	10.77	−14.91	−14.37	−14.08
10,000	−12.62	−18.53	14.33	8.42	−18.73	−18.32	−18.10
20,000	−14.74	−20.64	13.02	7.12	−20.82	−20.48	−20.30
40,000	−15.93	−21.83	12.27	6.37	−21.97	−21.70	−21.55
80,000	−16.62	−22.53	11.82	5.91	−22.64	−22.41	−22.29
160,000	−17.04	−22.95	11.53	5.63	−23.04	−22.84	−22.74
Cost to confine cattle	Yes	No	Yes	No	No	No	No
Centrifuge	Yes	Yes	No	No	Yes	Yes	Yes
Fertilizer credit	Yes	Yes	No	No	Yes	Yes	Yes
Interest rate	10%	10%	10%	10%	8%	12%	14%

[a] Modified Ashare et al. (1977) computer program utilizing a continuous-stirred tank; credit for electricity produced for on-farm use; 10% interest rate; 10% fuel escalation rate.

When one considers that, in most instances, it should be possible to finance digestion plants with approximately 25% of the cost borne by the owner and 75% financed at 1–2% above the prime lending rate, the potential profits are high. This is especially relevant because the owner is entitled to the full tax benefits (tax credits and depreciation). In some instances the owner of the plant may be a third-party investor who does not own the beef feedlot operation. In the instance of the 2500-head beef feedlot waste conversion facility, the total capital investment is $455,600 (Table XIV). For an investment of $113,900, the owner is entitled in the first year to a tax credit of $137,000 and a total ATB of $214,000. After 5 years, the total ATB is $766,000 on the original $113,900 investment. After 5 years, when the plant is fully depreciated, it might be advantageous for the third-party investor to sell the plant to the producer. Additional profit would be realized from the sale.

The same type of calculation can be made for the 10,000-head beef feedlot waste conversion by utilizing Tables XII and XV. In this instance the ATB after 5 years would be $4,362,000 on an investment of $253,250.

TABLE XIV

Economic Analysis of the Process to Produce Methane and Fertilizer from Beef Feedlot Waste on a 2500-Head Facility Operation[a]

Year	Gross revenues[b] ($1000)	O/M costs[c] ($1000)	Debt service[d] ($1000)	Cash flow[e] ($1000)	Interest ($1000)	Depreciation[f] ($1000)	Profit or loss[g] ($1000)	Tax credit ($1000)	Total ATB[h] ($1000)
1	566	366	80	119	49	72	(77)	137	214
2	622	384	80	157	46	106	(85)	0	111
3	685	403	80	200	43	101	(136)	0	126
4	753	424	80	248	39	101	(188)	0	146
5	829	445	80	303	35	101	(247)	0	169
6	911	467	80	363	30	0	(413)	0	140
7	1003	491	80	431	25	0	(486)	0	168
8	1103	515	80	507	20	0	(567)	0	200
9	1214	541	80	591	13	0	(658)	0	236
10	1335	568	80	686	73	0	(759)	0	276
11	1469	596	0	871	0	0	(871)	0	401
10-year total	9024	4608	806	3610	310	483	(3622)	137	1791
Internal rate of return = 222.2%									

[a] Modified Ashare et al. (1977) computer program utilizing a continuous-stirred tank digester; no cost to confine cattle; centrifuge; credit for fertilizer; credit for electricity produced for on-site use; 10% interest rate; 10% fuel escalation rate.
[b] Gross revenues: sale or credit for electricity, methane, and fertilizer.
[c] O/M costs: operation and maintenance costs.
[d] Debt service: cost of money (principal and interest).
[e] Cash flow: gross revenues − (operation and maintenance) − (cost of money).
[f] Depreciation: accelerated cost recovery at 15, 22, 21, 21, and 21%.
[g] Profit or loss: (operation and maintenance costs) + (interest) + (depreciation) − (gross revenues); parentheses indicate profit.
[h] Total after-tax benefits: (1 − tax rate) (profit or loss) + (tax credits) + (cash).

TABLE XV

Economic Analysis of the Process to Produce Methane and Fertilizer from Beef Feedlot Waste Contained on a 10,000-Head Facility Operation[a]

Year	Gross revenues[b] ($1000)	O/M costs[c] ($1000)	Debt service[d] ($1000)	Cash flow[e] ($1000)	Interest ($1000)	Depreciation[f] ($1000)	Profit or loss[g] ($1000)	Tax credit ($1000)	Total ATB[h] ($1000)
1	2265	883	180	1201	111	162	(1107)	308	911
2	2491	927	180	1383	104	238	(1221)	0	723
3	2741	973	180	1586	96	227	(1442)	0	806
4	3015	1023	180	1811	88	227	(1676)	0	906
5	3316	1074	180	2062	78	227	(1936)	0	1016
6	3648	1127	180	2340	68	0	(2451)	0	1015
7	4013	1184	180	2648	57	0	(2771)	0	1151
8	4414	1243	180	2990	44	0	(3126)	0	1302
9	4855	1305	180	3369	31	0	(3518)	0	1469
10	5341	1370	180	3790	16	0	(3954)	0	1654
11	5875	1439	0	4436	0	0	(4436)	0	2040
10 year totals	36,100	11,110	1800	23,180	697	1084	(23,207)	308	10,957

Internal rate of return = 605.9%

[a] Modified Ashare et al. (1977) computer program utilizing a continuous-stirred tank digester; no cost to confine cattle; centrifuge; credit for fertilizer; credit for electricity produced for on-site use; 10% interest rate; 10% fuel escalation rate.
[b] Gross revenues: sale or credit for electricity, methane, and fertilizer.
[c] O/M costs: operation and maintenance costs.
[d] Debt service: cost of money (principal and interest).
[e] Cash flow: gross revenues − (operation and maintenance) − (cost of money).
[f] Depreciation: accelerated cost recovery at 15, 22, 21, 21, and 21%.
[g] Profit or loss: (operation and maintenance costs) + (interest) + (depreciation) − (gross revenues); parentheses indicate profit.
[h] Total after-tax benefits: $(1 - \text{tax rate})$ (profit or loss) + (tax credits) + (cash).

X. Factors Affecting Profitability of Swine Waste Digestion

A. The Effect of Herd Size

If methane is the only vendable product, it is not economical to digest swine waste regardless of herd size. However, if swine waste is digested anaerobically, the methane is sold at $3/Mft^3, the digester solids are recovered and amended with nitrogen, phosphorus, and potassium to produce a 6-6-6 organic fertilizer and sold at $140/ton, and electricity is generated on site, the break-even point for a profitable digester operation is a herd size of 1305 animals (Table XVI). Most swine operations do not maintain that many animals on site or on a year-round basis (Overcash et al., 1983).

Additionally, no research has been conducted on the fertilizer value of digester swine waste residues. Undiluted swine waste cannot be spread on land for its fertilizer value at appreciable rates because it burns the crops.

XI. Factors Affecting Profitability of Poultry Waste Digestion

If methane is the only vendable product, it is not economical to digest poultry waste regardless of flock size. If all products, i.e.,

TABLE XVI

Effect of Number of Animals on Capital Requirements and Profitability of Swine Waste Conversion to Vendable Products[a]

Herd size	Total capital[b] investment ($1000)	Total plant investment ($1000)	Net CH_4[c] produced/year ($1000)	Net fertilizer[d] produced/year ($1000)	Average[e] after-tax benefits ($1000)
100	172.0	227.4	1.2	13.5	−105.9
200	215.1	284.5	2.5	27.1	−99.4
400	279.5	369.7	5.1	54.2	−84.3
800	375.8	497.1	10.3	108.5	−50.7
1600	522.3	690.8	20.7	217.0	20.7
3200	740.9	979.8	41.4	434.1	171.2
6400	1072.8	1418.8	82.8	868.2	482.4

[a] Modified Ashare et al. (1977) computer program utilizing a continuous-stirred tank digester; no capital to confine swine; credit for fertilizer; credit for electricity produced for on-farm use; 10% interest rate; 10% fuel escalation rate.
[b] With federal investment tax credits and energy tax credits.
[c] Methane value of $3/Mft^3.
[d] Organic 6-6-6 fertilizer value of $140/ton.
[e] Average of first 10 years.

TABLE XVII

EFFECT OF NUMBER OF ANIMALS ON CAPITAL REQUIREMENTS AND PROFITABILITY OF MESOPHILIC POULTRY WASTE CONVERSION TO VENDABLE PRODUCTS[a]

Flock size	Total capital[b] investment ($1000)	Total plant investment ($1000)	Net CH_4[c] produced/year ($1000)	Net fertilizer[d] produced/year ($1000)	Average[e] after-tax benefits ($1000)
31,250	214.5	283.7	3.1	1.3	−115.5
62,500	278.2	367.9	6.3	2.6	−116.4
125,000	373.5	493.9	12.7	5.3	−115.1
250,000	513.3	685.5	25.5	10.6	−108.0
500,000	736.5	974.1	51.1	21.3	−86.7
1,000,000	1066.8	1410.8	102.3	42.6	−33.8
2,000,000	1565.9	2070.9	204.7	85.2	87.9

[a] Modified Ashare et al. (1977) computer program utilizing a continuous-stirred tank digester; centrifuge; no capital to confine poultry; credit for fertilizer; credit for electricity produced for on-farm use; 10% interest rate; 10% fuel escalation rate.
[b] Federal investment tax credits and energy tax credits.
[c] Methane value of $3/Mft^3.
[d] Organic 6-6-6 fertilizer value of $140/ton.
[e] Average of first 10 years.

methane, fertilizer, and electricity, can be sold, then the break-even point for a profitable digester operation is 1,232,000 fryers under mesophilic conditions (Table XVII) and 1,000,000 fryers under thermophilic conditions (Table XVIII). Extensive studies have been performed using plastic bag digesters, which command a lower initial capital cost (Steinberger and Shih, 1984). Additionally, poultry waste has some value for refeed to chicken and cattle and has been used for that purpose. When the capital costs are factored into the computer program for plastic bag digesters, the break-even point for a profitable digester operation becomes 894,244 fryers under mesophilic conditions. This compares with 1,232,000 fryers using conventional digesters (Table XIX). Although there probably are some plastics that have properties which enable them to have a long life, the experience in our laboratory using dairy wastes and a plastic bag digester was not encouraging. After 1 year, at thermophilic conditions, the plasticizer leached from the plastic bags. The bags become brittle and burst. If plastic bags are to be used for the digester, it will be necessary to perform studies over long periods of time to test their durability.

TABLE XVIII

EFFECT OF NUMBERS OF ANIMALS ON CAPITAL REQUIREMENTS AND PROFITABILITY OF THERMOPHILIC POULTRY WASTE CONVERSION[a]

Flock size	Total capital[b] investment ($1000)	Total plant investment ($1000)	Net CH_4[c] produced/year ($1000)	Net fertilizer[d] produced/year ($1000)	Average[e] after-tax benefits ($1000)
31,250	169.9	224.7	3.1	1.3	−111.2
62,500	210.6	278.6	6.3	2.6	−109.9
125,000	271.1	358.5	12.7	5.2	−105.2
250,000	363.1	480.2	25.4	10.5	−93.1
500,000	501.3	663.0	50.7	21.1	−64.2
1,000,000	710.0	939.0	101.6	42.3	0
2,000,000	1025.3	1355.9	203.2	84.6	138.5

[a] Modified Ashare et al. (1977) computer program utilizing a continuous-stirred tank digester; centrifuge; no capital to confine poultry; credit for fertilizer; credit for electricity produced for on-farm use; 10% interest rate; 10% fuel escalation rate.
[b] Federal investment tax credits and energy tax credits.
[c] Methane value of $3/Mft^3.
[d] Organic 6-6-6 fertilizer value of $140/ton.
[e] Average of first 10 years.

XII. Factors Affecting Profitability of Water Hyacinth Digestion

A. TECHNICAL AND COST PARAMETERS SPECIFIC TO WATER HYACINTH SYSTEMS

A system that utilizes water hyacinths as a substrate for anaerobic digestion to methane must necessarily differ from a system that utilizes beef, swine, and poultry manure. Costs are incurred for pond construction ($25,000/acre), machinery for harvesting hyacinths ($180,000/37.33 acres), and operations (harvesting, grinding, and transport of hyacinths to the digester) ($2.50/wet ton). The digester system is common to any substrate system but the process conditions differ (Table VI).

In most water hyacinth systems either primary or secondary effluent is utilized as a source of nutrients for the production of hyacinths (Miller, 1983). In the economic analysis reported here, secondary effluent was utilized as a source of nutrients. Because hyacinths can efficiently remove nitrogen and phosphorus from wastewater, they have been used at the Iron Bridge Wastewater Treatment Plant in Orlando.

TABLE XIX

EFFECT OF DIGESTER TYPE ON THE CAPITAL NEEDS AND PROFITABILITY OF MESOPHILIC POULTRY MANURE CONVERSION[a]

	Conventional digester, continuous-stirred				Plastic bag digester			
Flock size	Total capital[b] investment ($1000)	Total plant investment ($1000)	Average cost of methane[c,d] ($/Mft^3)	Average[e] after-tax benefits ($1000)	Total capital[b] investment ($1000)	Total plant investment ($1000)	Average cost of methane[c,d] ($/Mft^3)	Average[e] after-tax benefits ($1000)
31,250	214.5	283.7	194.27	−115.5	140.8	186.2	181.89	−108.3
62,500	278.2	367.9	100.58	−116.4	156.9	207.5	90.38	−104.5
125,000	373.5	493.9	52.37	−115.1	198.6	262.7	45.03	−97.9
250,000	518.3	685.5	27.31	−108.0	271.4	358.9	22.13	−83.7
500,000	736.5	974.1	14.03	−86.7	379.7	502.2	10.29	−51.7
1,000,000	1066.8	1410.8	6.84	−33.8	570.4	754.4	4.23	14.9
2,000,000	1565.9	2070.9	2.81	87.9	931.0	1231.3	1.15	150.3

[a] Modified Ashare et al. (1977) computer program utilizing a centrifuge; no capital to confine poultry; credit for electricity produced for on-farm use; 10% interest rate; 10% fuel escalation rate.
[b] Federal investment tax credits and energy tax credits.
[c] Methane value of $3/Mft^3.
[d] Organic 6-6-6 fertilizer value of $140/ton.
[e] Average of first 10 years.

Florida, to remove nitrogen and phosphorus from secondary effluent, in place of chemical and microbiological methods to achieve the same purpose. A credit of $0.15/1000 gallons of secondary effluent treated via hyacinth ponds is the minimal credit that can be used for removal of nitrogen and phosphorus. This value was used in the calculations summarized here. Estimates of the credit range from $0.15 to $0.20/1000 gallons. A pond size of 2 acres is required to treat 100,000 gallons/day of secondary effluent. The 2 acres of hyacinth ponds will remove approximately 21 lb of nitrogen and 5 lb of phosphorus per day from 100,000 gallons of secondary effluent.

B. The Effect of Pond Size

If methane is the only vendable product and if the credit is $0.15/1000 gallons of secondary wastewater treated, it is not economical to digest hyacinths regardless of the size of the hyacinth ponds (Table XX). However, these facts do not preclude the use of hyacinths to treat wastewater and digestion of hyacinths to methane, because some municipalities may find it expedient to subsidize the process due to the

TABLE XX

Effect of Pond Size on Capital Requirements and Profitability of Water Hyacinth Conversion to Methane[a]

Pond size (acres)	Secondary waste water treated (10^6 gal/day)	Total capital[b] investment ($1000)	Total plant investment ($1000)	Net CH_4[c] produced/year ($1000)	Average[d] after-tax benefits ($1000)
1	0.5	356.4	758.8	3.8	−1409.5
25	1.25	453.5	1318.5	9.6	−1417.5
50	2.5	576.4	2199.8	19.2	−1427.1
100	5	768.1	3890.8	38.5	−1440.8
200	10	1071.6	7167.2	77.1	−1460.5
400	20	1554.5	13556.0	154.2	−1487.7
800	40	2333.8	26086.0	308.5	−1523.7
1000	50	2678.9	32293.0	385.6	−1537.3

[a] Modified Ashare et al. (1977) computer program utilizing a continuous-stirred tank digester; no capital for fertilizer; credit for electricity produced for on-site use; 10% interest rate; 10% fuel escalation rate.
[b] Federal investment tax credits and energy tax credits.
[c] Methane value of $3/Mft^3.
[d] Average of first 10 years.

relatively low capital requirements as compared to more conventional procedures. A 1000-acre pond that can treat 5×10^7 gallons/day of secondary wastewater would have a loss of 1537×10^6 per year (Table XX) or $0.084/1000 gallons of secondary wastewater treatment. This subsidy could be as low as $0.0342/1000 gallons depending on the alternative method for nitrogen and phosphorus removal. Additionally, the hyacinth digester residue has a value of at least $1.50/wet ton as compost material, which was not factored into the calculations reported here. For every 1000 gallons of wastewater treated, 0.106 wet ton of residue is produced with a value of $0.16. After costs for recovery and bagging of the hyacinth digester residue are subtracted from the value of the wet residue, no subsidy should be necessary; in fact, some profit should be realized.

If the same type of calculation is performed for various sizes of wastewater plants, the losses per 1000 gallons of secondary wastewater treatment with no recovery of solids are 1.25×10^6 gallons/day ($3.11/1000 gallons), 2.5×10^6 gallons/day ($1.56/1000 gallons), or 5×10^6 gallons/day ($0.79/1000 gallons).

If hyacinths are digested anaerobically, the methane is sold at $3/Mft^3, the digester solids are recovered and amended with nitrogen, phosphorus, and potassium to produce a 6-6-6 organic fertilizer sold at $140/ton, and secondary effluent is treated, the break-even point for a profitable digester is a pond size of 0.42 acres (Table XXI). The 0.42 acre could treat 21,000 gallons of wastewater per day. The value of the fertilizer makes the operation very profitable, especially as the sizes of the ponds increase (Table XXI). It should be emphasized that amended fertilizers have never been produced from hyacinth digester residues, even at pilot sites. The feasibility of the process must be proved and the marketability of the product established.

XIII. Effects of Potential Process Improvements on the Economics of Methane Production

A. Computer Model

One aim of this study was to examine the effects of possible improvements to the bioconversion process on the profitability of resource recovery. These improvements include reducing the concentration of refractory volatile solids increasing the methane bioconversion rate.

These particular studies were conducted in 1981 and used a computer program version and process input data (Varel et al., 1977) that

TABLE XXI

EFFECT OF POND SIZE ON CAPITAL REQUIREMENTS AND PROFITABILITY OF WATER HYACINTH CONVERSION TO VENDABLE PRODUCTS[a]

Pond size (acres)	Secondary waste water treated (10^6 gal/day)	Total capital[b] investment ($1000)	Total plant investment ($1000)	Net CH_4[c] produced/year ($1000)	Net fertilizer[d] produced/year ($1000)	Average[e] after-tax benefits ($1000)
10	0.5	439.6	868.5	3.8	749.6	342.9
25	1.25	564.5	1465.4	9.6	1874.1	1059.5
50	2.5	719.1	2388.6	19.3	3748.3	2259.0
100	5	955.6	4138.8	38.6	7496.6	4665.3
200	10	1322.5	7499.0	77.1	14,993.0	9488.2
400	20	1895.0	14,006.0	154.3	29,986.0	19,150.0
800	40	2801.1	26,704.0	308.4	59,973.0	38,497.0
1000	50	3197.2	32,978.0	385.6	74,966.0	48,176.0

[a] Modified Ashare et al. (1977) computer program utilizing a continuous-stirred tank digester; credit for electricity produced for on-site use; 10% interest rate; 10% fuel escalation rate.
[b] Federal investment tax credits and energy tax credits.
[c] Methane value of $3/Mft^3.
[d] Organic 6-6-6 fertilizer value of $140/ton.
[e] Average of first 10 years.

were accurate at that time (Wodzinski and Gennaro, 1981a,b). The program utilized capital cost data from the Hamilton Standard Installation at Kaplan Industries in Bartow, Florida. This model and the costs were accurate for relatively large installations.

The cost model calculations are reflected in the unit gas cost (UGC). The UGC is the cost, in dollars, to produce 1000 ft^3 of methane. This pricing structure is typically used in the petroleum industry.

The model differs from the model used in the preceding sections in that (1) it uses a different digester system, (2) it does not calculate after-tax benefits or third-party financing nor does it reflect current tax law, (3) it used 1981 capital costs and not 1985–1986 capital costs, and (4) the process conditions are different. This study is included here even though it was performed in 1981 because the conclusions are still accurate and relevant. The relative cost relationships exist now as they did in 1981. The actual numbers for unit gas cost may not express 1985–1986 economics, however, the calculations on effects of retention time, annual gas production, total solids degraded, and volatile solids degraded under the conditions studied should retain the accuracy they had in 1981.

B. Analysis of Factors Influencing Gas Production

Gas production was calculated by the equation

$$GT = 0.37*A(S0 - S1)*RNUM/HRT \tag{1}$$

where GT = gas production in liters gas/liters reactor/day, A = chemical oxygen demand to biodegradable influent solids, S0 = volatile solids concentration in influent (g/liter), S1 = volatile solids concentration in effluent (g/liter), RNUM = number of digesters, and HRT = hydraulic retention time (days). All of these values except S1 were provided in the data input either as one value or as several values using the brute force iteration.

S1 was calculated by the equation

$$S1 = ((1 - R)*S0/3(HRT*RK/RNUM) + 1)**RNUM) + R*S0 \tag{2}$$

where RK = first-order reaction rate constant (days^{-1}) and R = fraction of refractory volatile solids.

RK was calculated by the equation

$$RK = RK0*EXP(E/(T + 273)) \tag{3}$$

where RK0 = constant in Arrhenius rate expression, E = Arrhenius activation energy divided by gas constant, and T = temperature.

Gas production was dependent upon the amount of volatile solids converted. In order to increase gas production, the concentration of volatile solids in the effluent (S1) would have to be decreased. For any given temperature, retention time, and volatile solids influent concentration (S0), S1 can be decreased by decreasing the refractory volatile solids concentration (R) and/or increasing the Arrhenius rate constant (RK0). In this study, R was varied between 28 and 52%. RK0 was varied between 5.92×10^9 and 1.24×10^{11}. The analysis was performed using the brute force iteration. The temperature was maintained at 55°C. The herd number was held constant at 26,000. Total solids concentration in the influent was varied from 8 to 14% (w/w). Retention time was varied from 1 to 10 days. A modification of the output from the iteration was made. S1 and the percentage volatile solids conversion (VSCON) were calculated instead of the heat requirements and makeup water requirements. The analysis was performed with and without the fertilizer plant option.

Extensive documentation of the logic flow, available options, process conditions, and capital cost information is available in Ashare et al. (1977) and Wodzinski and Gennaro (1981b).

C. Effect of Retention Time on Volatile Solids Conversion and Annual Gas Production

The percentage volatile solids conversion and annual gas production increased as retention time increased (Figs. 3 and 4). The rate of increase was greater at short retention times than at long retention times.

D. Effect of Total Solids Concentration on Unit Gas Cost

Unit gas cost decreased as total solids concentration increased (Fig. 5). If methane is the desired product, it is beneficial to increase the total solids concentration as high as possible.

E. Effect of Refractory Solids Concentration on Percentage Volatile Solids Conversion and Gas Production

The percentage volatile solids and total gas production increased as refractory volatile solids concentration decreased (Figs. 6 and 7). Volatile solids conversion was 16.7% when the refractory volatile solids

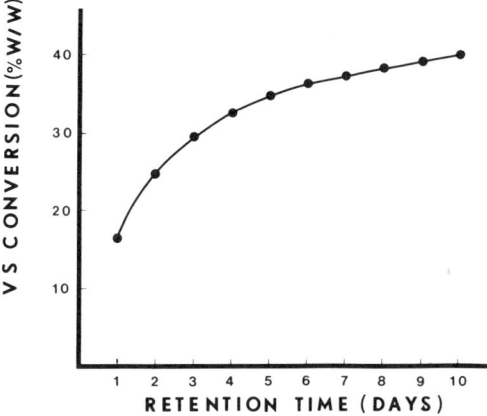

FIG. 3. Effect of retention time on percentage volatile solids (VS) conversion. (●), R = 52.4, RK0 = 5.92 × 10^9.

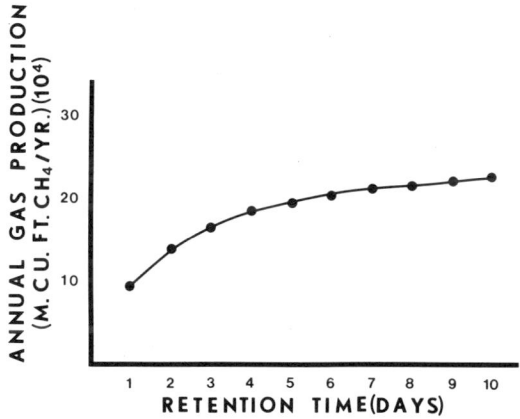

FIG. 4. Effect of retention time on annual gas production. (●), R = 52.4, RK0 = 5.92 × 10^9.

concentration was 52% at a 1-day retention time. The percentage conversion was highest when the retention time was 10 days and the refractory volatile solids concentration was 28%. This condition resulted in 60.6% volatile solids conversion. Gas production was lowest and highest under these two conditions. The first condition resulted in 9.357 × 10^4 Mft^3 methane/year, and the second resulted in 3.433 × 10^5 Mft^3 methane/year.

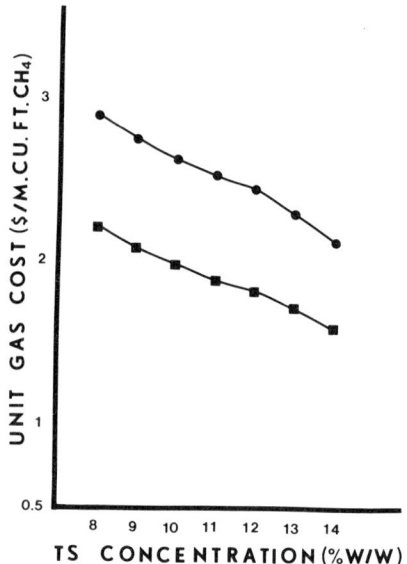

FIG. 5. Effect of total solids (TS) concentration on unit gas cost. (●), HRT = 1; (■), HRT = 3.

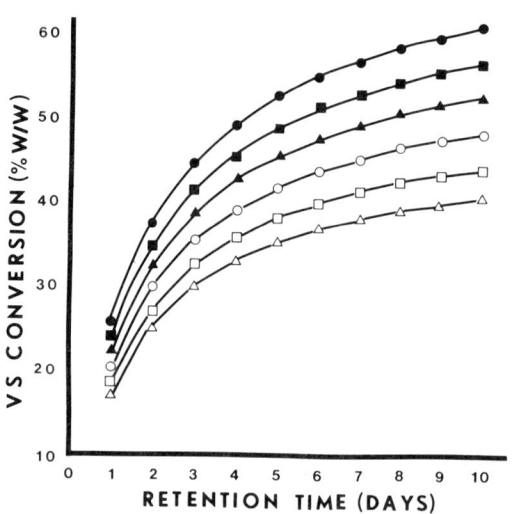

FIG. 6. Effect of refractory volatile solids (VS) concentration on percentage volatile solids conversion. (●), R = 28%; (■), R = 33%; (▲), R = 38%; (○), R = 43%; (□), R = 48%; (△), R = 52%.

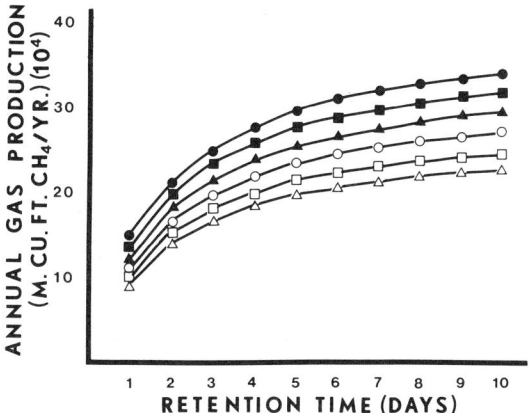

FIG. 7. Effect of refractory volatile solids concentration on annual gas production. (●), $R = 28\%$; (■), $R = 33\%$; (▲), $R = 38\%$; (○), $R = 43\%$; (□), $R = 48\%$; (△), $R = 52\%$.

F. Effect of Reaction Rate Constant on Percentage Volatile Solids Conversion and Gas Production

Percentage volatile solids conversion and gas production increased as the reaction rate constant (RK0) increased (Figs. 8 and 9). The lowest values for conversion and gas production were obtained at a 1-day retention time with RK0 equal to 5.92×10^9. Those values were 16.7% and 9.436×10^4 Mft³ methane/year. Highest values of 47.6% and 2.695×10^5 Mft³ methane/year were obtained on a 10-day retention time when RK0 = 1.24×10^{11}.

G. Effect of Refractory Volatile Solids Concentration on Unit Gas Cost without a Fertilizer Plant

The unit gas cost decreased as the refractory volatile solids concentration decreased (Fig. 10). The highest unit gas cost was $32.75/Mft³ methane at 52% refractory volatile solids, a 1-day retention time, and 8.0% total solids. The lowest unit gas cost was $9.74/Mft³ methane at 28% refractory volatile solids, a 10-day retention time, and 14% total solids. Unit gas cost increased as total solids decreased.

H. Effect of Refractory Volatile Solids Concentration on Unit Gas Cost with a Fertilizer Plant

Unit gas cost decreased as refractory volatile solids concentration decreased (Fig. 11). The unit gas cost does not, however, continue to

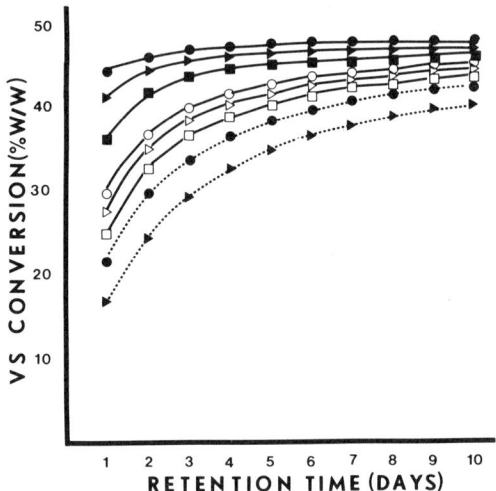

FIG. 8. Effect of reaction rate constant on percentage volatile solids (VS) conversion. (●), RK0 = 1.24×10^{11}; (▲), RK0 = 6.51×10^{10}; (■), RK0 = 3.55×10^{10}; (○), RK0 = 1.78×10^{10}; (△), RK0 = 1.48×10^{10}; (□), RK0 = 1.18×10^{10}; (●--●), RK0 = 8.88×10^{9}; (▲--▲), RK0 = 5.92×10^{9}.

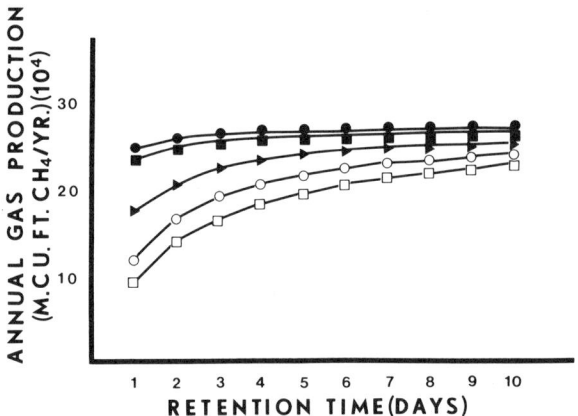

FIG. 9. Effect of reaction rate constant on annual gas production. (●), RK0 = 1.24×10^{11}; (■), RK0 = 6.51×10^{10}; (▲), RK0 = 1.78×10^{10}; (○), RK0 = 8.88×10^{9}; (□), RK0 = 5.92×10^{9}.

decline with increased retention time when the refractory volatile solids concentration is greater than 33% at 10% total solids. Unit gas cost does decrease as influent total solids increase. The highest unit gas cost was $9.23/Mft³ methane at 8% total solids, a 1-day retention time,

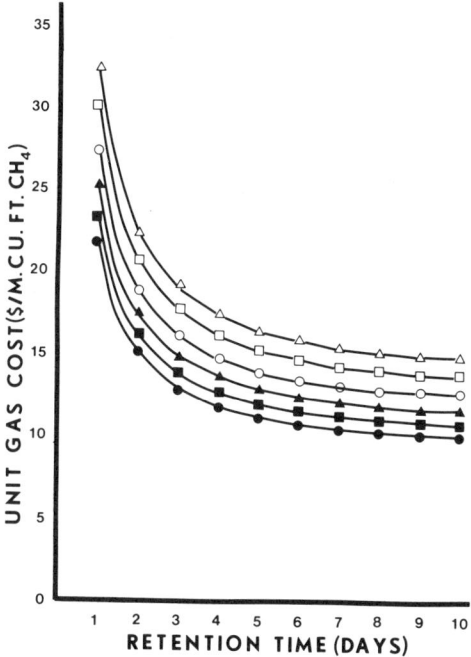

FIG. 10. Effect of refractory volatile solids concentration on unit gas cost without a fertilizer plant. (●), R = 28%; (■), R = 33%; (▲), R = 38%; (○), R = 43%; (□), R = 48%; (△), R = 52%.

and 28% refractory volatile solids. The lowest unit gas cost was $1.58/Mft^3 methane at 14% total solids, a 4-day retention time, and 52% refractory volatile solids. The lowest unit gas cost at 10% total solids and 52% refractory volatile solids was at a 4-day retention time. The unit gas cost was $1.96. However, the unit gas costs at 3- and 5-day retention times were $1.97 and $1.98, respectively. The lowest unit gas costs at 48% refractory volatile solids and 10% total solids were at 4- to 7-day retention times with costs of $2.68, $2.64, $2.65, and $2.67, respectively.

I. Effect of Reaction Rate Constant on Unit Gas Cost without a Fertilizer Plant

Unit gas cost (UGC) decreased as the reaction rate constant (RK0) increased (Fig. 12). The unit gas cost did not continue to decline with retention time when the reaction rate constant was greater than 1.78×10^{10} at 10% total solids and a refractory volatile solids concentration of

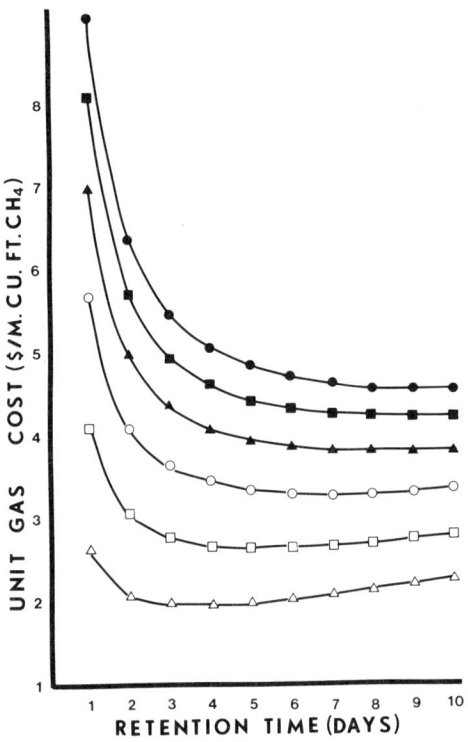

Fig. 11. Effect of refractory volatile solids concentration on unit gas cost with a fertilizer plant. (●), R = 28%; (■), R = 33%; (▲), R = 38%; (○), R = 43%; (□), R = 48%; (△), R = 52%.

52%. The highest UGC was $32.75/Mft^3 methane at 8.0% total solids, a 1-day retention time, and when the reaction rate constant was 5.92×10^9. The lowest UGC of $12.06 occurred at 13% of total solids, a 3-day retention time, and when RK0 was 1.24×10^{11}. The lowest UGC at 10% total solids was $12.22 when the retention time was 3 days and RK0 was 1.24×10^{11}. When the retention time was 4 and 5 days, the UGC was $12.25 and $12.30, respectively.

J. Effect of Reaction Rate Constant on Unit Gas Cost with a Fertilizer Plant

Unit gas cost decreased as the reaction rate constant (RK0) increased (Fig. 13). The UGC did not continue to decrease with retention time.

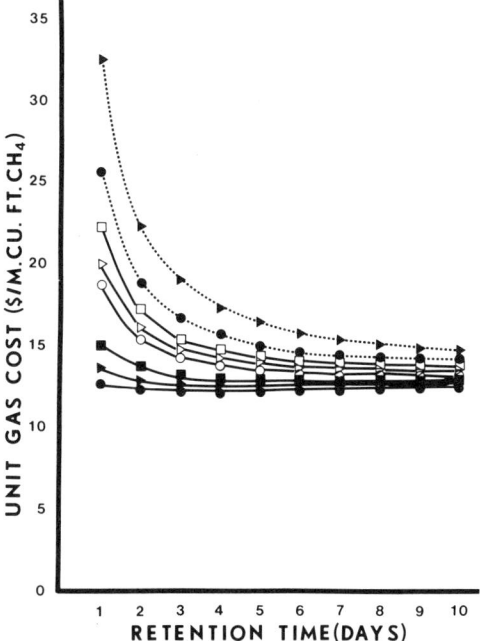

FIG. 12. Effect of reaction rate constant on unit gas cost without a fertilizer plant. (●), RK0 = 1.24×10^{11}; (▲), RK0 = 6.51×10^{10}; (■), RK0 = 3.55×10^{10}; (○), RK0 = 1.78×10^{10}; (△), RK0 = 1.48×10^{10}; (□), RK0 = 1.18×10^{10}; (●--●), RK0 = 8.88×10^{9}; (▲--▲), RK0 = 5.92×10^{9}.

When RK0 was equal to 5.92×10^9, the UGC decreased to $1.96 at a 4-day retention time. After 4 days, the UGC increased. The point of inflection occurred at shorter retention times as RK0 was increased. The highest unit gas cost of $2.88 occurred when RK0 was equal to 5.92×10^9, the retention time was 1 day, and the total solids concentration was 8%. The lowest unit gas cost was $0.90 at a 1-day retention time, 14% total solids, and when RK0 was equal to 1.24×10^{11}.

K. Effect of Refractory Volatile Solids Concentration and Reaction Rate Constant on Unit Gas Cost without a Fertilizer Plant

Unit gas cost decreased as refractory volatile solids decreased and as the reaction rate constant increased (Fig. 14). The greatest effect on decreasing the unit gas cost was the concentration of refractory volatile solids. However, the unit gas cost was never below $8/Mft^3 methane.

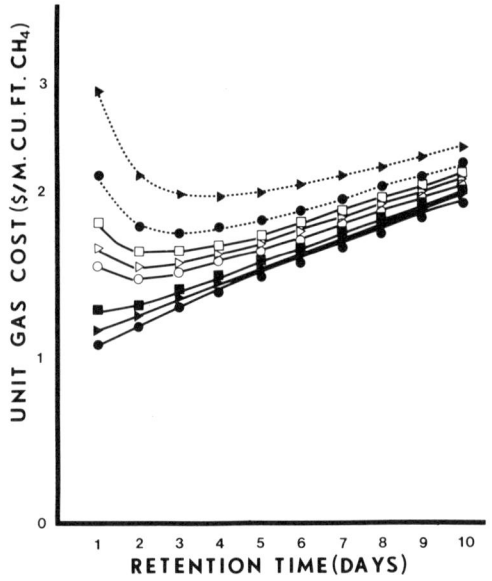

FIG. 13. Effect of reaction rate constant on unit gas cost with a fertilizer plant. (●), RK0 = 1.24 × 10^{11}; (▲), RK0 = 6.51 × 10^{10}; (■), RK0 = 3.55 × 10^{10}; (○), RK0 = 1.78 × 10^{10}; (△), RK0 = 1.48 × 10^{10}; (□), RK0 = 1.18 × 10^{10}; (●--●), RK0 = 8.88 × 10^{9}; (▲--▲), RK0 = 5.92 × 10^{9}.

L. Effect of Refractory Volatile Solids Concentration and Reaction Rate Constant on Unit Gas Cost with a Fertilizer Plant

Unit gas cost decreased as the refractory volatile solids concentration and the reaction rate constant increased (Fig. 15). The greatest effect on decreasing the unit gas cost was increasing the concentration of refractory volatile solids. The unit gas costs were consistently lower than $4/Mft^3. The unit gas costs were lower than $2.15 when the refractory volatile solids concentration was 52%.

M. Discussion

The effects of refractory volatile solids concentration and reaction rate constant on the economics of bioconversion of dairy waste were studied. The refractory volatile solids concentrations were varied between 28 and 52%. Varel et al. (1977) reported that 52.4% of the volatile solids in dairy manure is refractory. Analysis of dairy cattle manure has

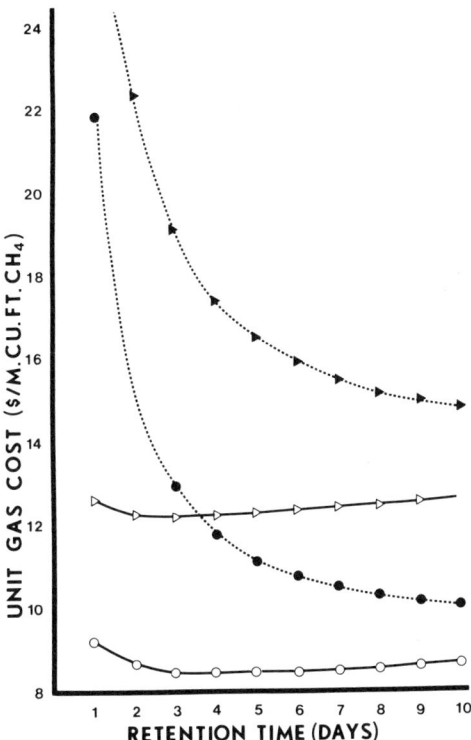

FIG. 14. Effect of refractory volatile solids concentration and reaction rate constant on unit gas cost without a fertilizer plant. (▲), RK0 = 5.92 × 10^9, R = 52%; (●), RK0 = 5.92 × 10^9, R = 28%; (△), RK0 = 1.24 × 10^{11}, R = 52%; (○), RK0 = 1.24 × 10^{11}, R = 28%.

shown that lignin comprises 28% of the volatile solids (Varel et al., 1977). It is unlikely that any significant portion of the lignin would be degraded within 10 days at 55°C. This information was used to determine the parameters of this study.

The reaction rate constant was varied between 5.92 × 10^9 and 1.24 × 10^{11}. The value of 5.92 × 10^9 was determined by Ashare et al. (1977) through a least-squares fit of experimental data obtained by several investigators. This was the minimum attainable rate. The higher values that were used were estimates of potential increases in rate that might be obtained if the fermentation was optimized by strain selection, physical and chemical parameters, and nutritional techniques. Aerobic fermentations have been optimized by a combination of these techniques. Rates of fermentation and increases in total yields of

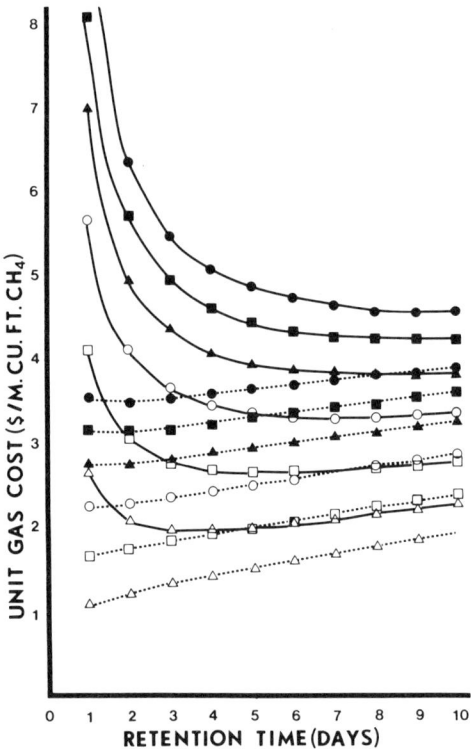

FIG. 15. Effect of refractory volatile solids concentration and reaction rate constant on unit gas cost with a fertilizer plant. (●), R = 28%; (■), R = 33%; (▲), R = 38%; (○), R = 43%; (□), R = 48%; (△), R = 52%; (———), RK0 = 5.92 × 10^9; (--------), RK0 = 1.24 × 10^{11}.

desired products are commonplace. The values used were 1.5, 2.0, 2.5, 3.0, 6.0, 11.0, and 20.9 times the rate calculated by Ashare.

Higher yields of methane should be produced by a digestion process that degrades a higher proportion of the volatile solids in the feed material. Yields might also be increased by a digestion system that contained organisms that would convert the volatile solids to methane at a faster rate.

Conversion of dairy wastes to methane is economical if the solids are recovered from the digester effluent and are upgraded to a 6-6-6 fertilizer before sale. Without this type of solids recovery, the process is uneconomical at any of the reaction rate constants and refractory volatile solids concentrations studied. The lowest unit gas cost without a fertilizer plant reported in this study was more than $8/Mft³ methane.

The wholesale price of methane is presently $3/Mft^3. The unit gas cost was lower than these figures when the fertilizer plant was included in the calculations.

At the present digestion conditions of 52% refractory volatile solids and a reaction rate constant of 5.92×10^9, the unit gas cost with a fertilizer plant was $1.98/Mft^3 methane on a 5-day retention time. Hashimoto et al. (1979) have reported a unit gas cost of $2.81/Mft^3 methane for an installation of 50,000 head of beef cattle. The calculations did not utilize credits for carbon dioxide production and do not reflect a credit for refeeding of solids. The unit gas cost was calculated at $2.47/Mft^3 methane when a credit of $60.96/ton was realized from refeed of solids. The refeed credit does not require an upgrading cost before sale. The only cost necessary is a recovery cost. This is also necessary in fertilizer production. The cost of recovery of the solids for fertilizer may be less than the cost of solids for refeed. Hashimoto's costs were based upon results obtained with a 5,700-liter plant digester and recalculated to a 50,000-head installation.

Decreasing the refractory volatile solids concentration increases methane production but it also increases the unit gas cost if a fertilizer plant is used. The solids have a greater value in their present form than as methane. The unit gas cost can be reduced by increasing the reaction rate. This will increase the amount of methane produced and will decrease the required digester volume. A lower volume will result in a lower capital cost expenditure. The decreases in unit gas cost that result from an increased rate are attractive from an economic standpoint. At the present reaction rate of 5.92×10^9, the process is economical. Any increase in rate will result in additional profits for the operator of such a system.

The digestion process should be optimized for maximum methane production, in the shortest time, with minimum solids degradation. Further experimentation is needed to determine the minimum digestion time for stabilization of the dairy waste solids and removal of substances that are toxic to plants, if any. Further study is also needed to determine the advantages of fertilizer with a highly organic base and low leachability, such as dairy cattle waste. The effect of digested wastes on upgrading poor soils such as those found in marginal farm lands should be examined. Not only could this result in increased crop acreage, but it might also allow the product to command a higher price than inorganic fertilizer. These studies also illustrate that determination of economic impacts on processes can be used to assess whether specific applied research projects will yield economic benefits if the premises of the research project are brought to a successful conclusion.

At the very least, they might be used in conjunction with technical considerations to establish research priorities for applied projects.

XIV. Conclusions

It is not profitable to anaerobically digest water hyacinths or dairy, beef feedlot, swine, or poultry wastes if methane is the only product. It becomes profitable only if credits toward the process can be accrued from the sale of other products or from pollution abatement.

Economic analyses indicate that anaerobic digestion of water hyacinths or animal wastes would be very profitable for large installations if (1) the methane is sold, (2) electricity is generated on site from the methane and sold to power companies, and (3) and the digester solids are recovered and amended with nitrogen, phosphorus, and potassium to produce a slowly leachable organic fertilizer. Under these conditions, the break-even point for profitability is the bioconversion of the waste from 948 dairy cows, 998 beef cattle, 1305 swine, and 1,132,000 chickens.

The number of animals required for the break-even point for profitability is considerably higher than claimed in previously published reports (Jones and Ogden, 1985; Pike et al., 1979). Some of the increases are due to changing economic conditions. The bulk of the increase is due to differences in opinion on the value of credits for the products of anaerobic digestion and the capital costs for reliable equipment.

Water hyacinths produced in a 0.42-acre pond that was charged daily with 21,000 gallons of secondary effluent, if digested, would be a profitable operation if fertilizer was made from the digester solids, methane was sold, and a credit of $0.15/100,000 gallons of wastewater treated was applied. It might be feasible for municipalities to subsidize large water hyacinth installations (400+ acres) that would treat 2.0×10^7 gallons of secondary effluent for the removal of nitrogen and phosphorus. The subsidy would be low on the basis of a per 1000-gallons wastewater treated. The subsidies would be feasible if the municipality possessed the land for the ponds and if capital was at a premium, since alternative methods for nitrogen and phosphorus production are more capital intensive.

Widespread implementation of the anaerobic digestion of animal wastes or water hyacinths to methane and other vendable products is contingent on future basic and developmental research. Pilot-scale demonstrations of the production of slowly leachable amended fertilizers from digester solids should be performed. If successful, the marketability of the product should be verified.

Research is needed to determine reactor conditions that produce methane at maximum rates with the minimum amount of solids destruction; stabilized solids produce a more desirable fertilizer. The profitability of anaerobic digestion would be increased if strains of microorganisms could be established that have faster reaction rates for the production of methane. The profitability of anaerobic digestion will be impacted negatively if changes in tax laws eliminate energy or investment tax credits or if depreciation periods are lengthened.

The present temporary supply of cheap energy has dampened enthusiasm for the production of methane by anaerobic digestion. Interest will be rekindled when the oil glut subsides. This hiatus should be used to discover and develop other potential products from the anaerobic digester solids.

ACKNOWLEDGMENT

This work was supported by the State of Florida, Coordinating Council on the Restoration of the Kissimmee River Valley and Taylor Creek–Nubbin Slough Basin, a Non-Instructional Computer Grant at the University of Central Florida, and a Travel Grant from the Department of Biology at Memphis State University.

NOTE ADDED IN PROOF

On September 27, 1986, the Senate passed the Tax Reform Act of 1986, which made sweeping changes in the tax law. These changes have a direct effect on the economics reported in this communication. The new Tax Act abolished the 10% investment tax credit and decreased the federal tax credit for alternative energy products utilizing biomass from 15% to 10% for January 1, 1987, to December 31, 1987. After December 31, 1987, the biomass energy credit will be abolished. The Tax Reform Act of 1986 also abolished the accelerated cost recovery system (ACRS), which permitted depreciation of the capital cost of the digester system over a 5-year period at set percentages per year. It replaced the ACRS with a schedule of depreciation for "machinery" over a 7-year period and permits the use of a 200% declining balance method. The maximum tax rate for corporations was reduced from 46% to 34% and for individuals from 50% to 28%.

We have attempted to quantify the effects of these tax law changes on the economics of the bioconversion of biomass presented here. Preliminary estimates based on changes reported in *The Price Waterhouse Guide to the New Tax Law* (Anonymous, 1986) indicate that the profitability of the water hyacinth system in the 10-year after-tax benefits (ATB) totals presented in Table IV will be reduced in 1987 by ~14%, or $483,000. Approximately $129,000 of the reduction is due to the abolishment of the investment credit and the decrease in the energy credit. The rest of the reduction is due to changes in the depreciation schedule and especially due to the reductions in the tax rate. It should be emphasized that ATBs provided tax shelters for third-party financers who had profits on which they were paying 46–50% taxes. The equation for the calculation of ATBs is in the footnote to Table XIV. "Profit" in third-party financing is listed as a negative value since it detracts from the aim of the shelter. As the tax rate goes down, the ATBs to the third-party financer are decreased, especially in operations that are marginally profitable when calculated by conventional techniques. The profitability of the process to the third-party financer will be reduced further in 1988 when the energy credit is completely abolished. In our example, the profit will be $86,900 less than in 1987, or a total of 31% less profit in 1988 than in 1986.

Similar effects of the new tax law should be operative for the ATBs listed for the dairy, steer, swine, and poultry scenarios. The percentages listed above should be comparable at the same level of ATBs regardless of the type of operation. However, the effect diminishes with increased profitability of larger operations. The digester cost is raised to the 0.6 power as the size of the operation doubles. Therefore, the reduction in the federal tax credits plays a proportionally diminishing role as the size of the installation increases.

In the Conclusions section, we estimated that the break-even point for profitability was: 984 dairy cows, 998 beef, 1305 swine, and 1,132,000 poultry. We estimate that these break-even points should be increased by at least 14% for 1987 and 31% in 1988 to compensate for the changes in the Tax Reform Act of 1986.

REFERENCES

Ashare, E., Wise, D. L., and Wentworth, R. L. (1977). "Fuel Gas Production from Animal Residues," pp. 1–202. Dynatech Report No. 1551, ERDA, Coo/2991-10.

Brenner, C. W. (1983). "Third Party Financing: A Primer for the Baffled Energy Professional," pp. 1–128. Commonwealth Energy Group, Winchester, Massachusetts.

Canton, G. H., Buffington, D. E., Collier, R. J., and Thatcher, W. W. (1979). In "Methane and Other Vendable Products from Animal Wastes" (P. M. McCaffrey, T. Beemer, and S. E. Gatewood, eds.), pp.175–183. Coordinating Council on the Restoration of the Taylor Creek–Nubbin Slough Basin, Tallahassee, Florida.

Cheynoweth, D. P., Dolenc, D. A., Ghosh, S., Henry, M. P., Jerger, D. E., and Srivastava, V. J. (1982). *Biotech. Bioeng. Symp.* **12**, 381–398.

Coppinger, E., Baylor, D., and Smith, K. (1978). "Report on the Design and First Year Operation of a 50,000 Gallon Anaerobic Digester at the State Honor Farm, Monroe, Washington," pp. 1–88. DOE, Div. Solar Tech., Fuels from Biomass Branch, EG-77-C-06-1016.
Fischer, J. R., Iannotti, E. L., Porters, J. H., and Garcia, A. (1979). *Trans. Am. Soc. Agric. Eng.* **22**, 370–374.
Hashimoto, A. G. (1982). *Biotech. Bioeng.* **24**, 9–23.
Hashimoto, A. G., Chen, Y. R., and Prior, R. L. (1979). *J. Soil Water Conserv.* 34, 16–19.
Hinley, J. V. (1979). *In* "Methane and Other Vendable Products from Animal Wastes" (P. M. McCaffrey, T. Beemer, and S. E. Gatewood, eds.), pp. 150–154. Coordinating Council on the Restoration of the Kissimmee River Valley and Taylor Creek–Nubbin Slough Basin, Tallahassee, Florida.
Jewell, W. J., Davis, H. R., Gunkel, W. W., Lathwell, D. J., Martin, D. J., Jr., McCarty, T. R., Morris, G. R., Price, O. R., and Williams, D. W. (1976). "Bioconversion of Agricultural Wastes for Pollution Control and Energy Conversion," pp. 1–321. ERDA, Div. Solar Energy, NSF, RANN. TID-27164.
Jewell, W. J., Capener, H. R., Dell'Orto, S., Fanfoni, K. J., Hayes, T. D., Leuschner, A. P., Miller, T. L., Sherman, D. F., Van Soest, P. J., Wolin, M. J., and Wujcik, W. J. (1978). "Anaerobic Fermentation of Agricultural Residue: Potential for Improvement and Implementation. Final Report," pp. 1–426. DOE, Div. Solar Tech., EY-76-S-02-2981.
Jones, Jr., H. B., and Ogden, E. A. (1985). *South. Biomass Energy Res. Conf., 3rd, Gainesville, Florida* p. 20.
Miller, G. R. (1983). "An Economic and Systems Assessment of the Concept of a Water Hyacinth Wastewater Treatment/Methane Production System," pp. 1–57. Gas Research Institute, GRI-82/0066.
Overcash, M. R., Humenik, F. J., and Miner, J. R. (1983). "Livestock Waste Management," Vols. 1 and 2. CRC Press, Boca Raton, Florida.
Pike, R. W., Kebanal, E. S., and Ngkwingking, J. (1979). "Feasibility Study in the Conversion of Animal Feedlot Wastes to Useful Energy: Review of the Economics of Methane Production from Animal Wastes," pp. 1–56. DOE, EM-78-GOL-5261.
Rodriguez, L. (1979). *In* "Methane and Other Vendable Products from Animal Wastes" (P. M. McCaffrey, T. Beemer, and S. E. Gatewood, eds.), pp. 158–161. Coordinating Council on the Restoration of the Kissimmee River Valley and Taylor Creek–Nubbin Slough Basin, Tallahassee, Florida.
Sartain, J. B. (1979). *In* "Methane and Other Vendable Products from Animal Wastes" (P. M. McCaffrey, T. Beemer, and S. E. Gatewood, eds.), pp. 162–174. Coordinating Council on the Restoration of the Kissimmee River Valley and Taylor Creek–Nubbin Slough Basin, Tallahassee, Florida.
Steinberger, S. C., and Shih, J. C. H. (1984). *Biotech. Bioeng.* **26**, 537–543.
Varel, V. H., Issacson, H. R., and Bryant, M. P. (1977). *Appl. Environ. Microbiol.* **33**, 298–307.
Varel, V. H., Hashimoto, A. G., and Chen, Y. R. (1980). *Appl. Environ. Microbiol.* **40**, 217–222.
Wodzinski, R. J., and Gennaro, R. N. (1981a). "Initial Feasibility Assessment: Resource Recovery Potential of Dairy Wastes in Okeechobee, Florida," pp. 1–64. Coordinating Council on the Restoration of the Kissimmee River Valley and Taylor Creek–Nubbin Slough Basin, Tallahassee, Florida.
Wodzinski, R. J., and Gennaro, R. N. (1981b). "Initial Feasibility Assessment: Process Cost Analysis for Dairy Wastes Resource Recovery in Okeechobee, Florida," pp. 1–25. Coordinating Council on the Restoration of the Taylor Creek–Nubbin Slough Basin, Tallahassee, Florida.

Wodzinski, R. J., and Gennaro, R. N. (1979). *In* "Methane and Other Vendable Products from Animal Wastes" (P. M. McCaffrey, T. Beemer, and S. E. Gatewood, eds.), pp. 126–149. Coordinating Council on the Restoration of the Kissimmee River Valley and Taylor Creek–Nubbin Slough Basin, Tallahassee, Florida.

The Microbial Production of 2,3-Butanediol

ROBERT J. MAGEE* AND NAIM KOSARIC†

*Institute for Research in Construction
National Research Council
Ottawa, Ontario, Canada K1A 0R6

†Chemical and Biochemical Engineering
Faculty of Engineering Science
The University of Western Ontario
London, Ontario, Canada N6A 5B9

I. Introduction

Microbial production of 2,3-butanediol was investigated in 1906 by Harden and Walpole and in 1912 by Harden and Norris. The bacterium employed in these early studies was *Klebsiella pneumoniae*. Donker (1926) is credited with the initial observation of diol accumulation in cultures of *Bacillus polymyxa*, whereas industrial-scale formation of this compound is believed to have been first proposed by Fulmer et al. (1933). Shortages of the strategic compound 1,3-butadiene during World War II stimulated intense research efforts in the formation of diol, culminating in the development of pilot-scale operations for both its manufacture and its conversion to butadiene.

Also known as 2,3-butylene glycol, 2,3-dihydroxybutane, or dimethylene glycol, 2,3-butanediol is a colorless and odorless liquid. Possessing a high boiling point (approximately 180°C), this compound is not easily recovered by conventional distillation. The low freezing point of the levo isomer (−60°C) is the basis of the commercial interest in its use as an antifreeze agent (Clendenning, 1946; Clendenning and Wright, 1946). At 35°C, the viscosity of the levo isomer (0.0218 Pa·s) is approximately one-third that of the meso form.

Commercial application of 2,3-butanediol is not limited to the manufacture of butadiene or antifreeze. With a heating value of 27,200 kJ/kg (Flickinger, 1980), diol compares favorably with methanol (22,100 kJ/kg) and ethanol (29,100 kJ/kg) for use as a liquid fuel. An equimolar mixture of butanediol and ethanol (formed as a by-product in microbial fermentations) has an energy content of 27,700 kJ/kg. Condensation of diol to methyl ethyl ketone (MEK) coupled with subsequent hydrogenation yields octane isomers that can be used to produce high-quality aviation fuels. MEK can also be used as a solvent for resins and laquers (Villet, 1981). Esterification of butanediol forms precursors of polyurethane foams. Diacetyl, formed by catalytic dehydrogenation of the diol, is a highly valued food additive. Butanediol has also been shown to have potential application in the manufacture of printing inks, perfumes, fumigants, moistening and softening agents, explosives and plasticizers, and as a carrier for pharmaceuticals.

Two reviews have, in the past, dealt with the formation of butanediol by microorganisms (Ledingham and Neish, 1954; Long and Patrick, 1963), whereas summaries of certain aspects have been included in more recent articles (Bu'Lock, 1975; Rosenberg, 1980; Jansen and Tsao, 1983). It is the object of this paper to provide a comprehensive survey of diol production, including biochemistry, microbiology, and process engineering.

II. Microbiology

Although diol formation has been observed in several yeasts (Kluyver and Donker, 1925; Fields and Richmond; 1967), yields are extremely low. Thus, bacteria are at present the only organisms to be considered of industrial importance in the production of butanediol. Key species are found in the genera *Klebsiella, Bacillus, Serratia*, and *Pseudomonas*. The wide distribution among bacteria of the ability to generate this compound is evident from the classification of key diol producers (Fig. 1). Each of these organisms will be discussed individually in terms of the major characteristics of the fermentations they conduct. Table I summarizes typical product yields and conversion efficiencies.

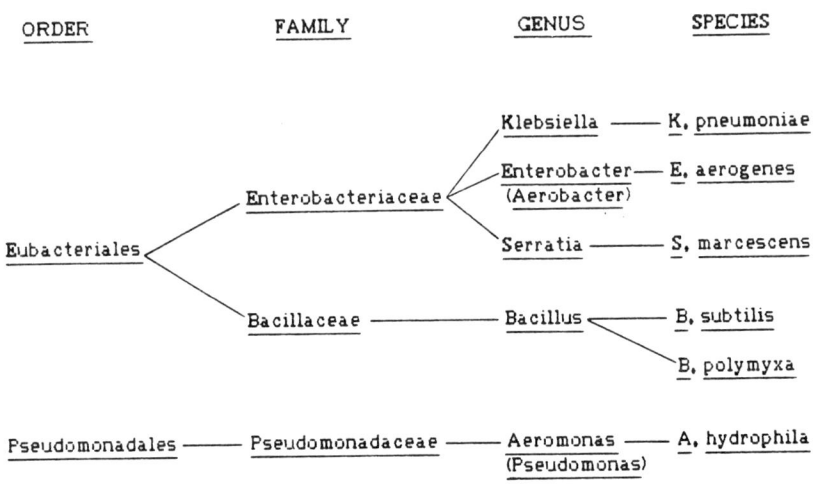

FIG. 1. Classification of major butanediol-producing bacteria (source, Buchanan, 1974).

TABLE I

GLUCOSE DISSIMILATION BY MAJOR DIOL-PRODUCING SPECIES[a]

Organism	Glucose		Aeration	Solvent yield (g/liter)					AMC + diol		Ref.[d]
	g/liter	%Used		Diol	AMC	A + BD	EtOH	HAc	%Theory	g/liter/hr	
K. pneumoniae	20	100(96)	Anaerobic	7.39	0.09	7.48	2.69	0.20	74.8	0.08	1
K. pneumoniae	10	100(24)	Aerobic	0.20	Nil	0.20	0.31	0.40	4.0	0.01	2
K. pneumoniae	10	100(24)	Anaerobic	3.45	Nil	3.45	1.76	0.13	69.0	0.14	2
B. polymyxa	20	99.5	Aerobic	4.82	0.62	5.45	2.25	NR	54.1	NR	3
B. ploymyxa	20	98.0	Anaerobic	4.83	0.17	5.01	3.42	NR	49.0	NR	3
B. subtilis	50	53(168)	Aerobic	4.06	3.22	7.28	Nil	0.38	29.1	0.04	4
A. hydrophila	[b]	[c]	Anaerobic	6.01	0.19	6.20	2.92	0.34	—	—	5
S. marcescens	50	100(120)	Aerobic	15.63	0.81	16.44	4.90	0.58	65.7	0.14	6
S. marcescens	50	93(408)	Anaerobic	13.50	0.06	13.56	4.85	Nil	54.2	0.03	6

[a] Numbers in parentheses represent time at which observations were recorded (hours); NR, not reported.
[b] 10–20 g/liter.
[c] 5.49 g consumed in 3–4 days.
[d] References: (1), Mickelson and Werman (1938); (2), Yu and Saddler (1982a); (3), Stahly and Werkman (1942); (4), Blackwood et al., (1947); (5), Stanier and Adams (1944); and (6), Neish et al. (1947).

Three stereoisomeric forms of 2,3-butanediol may be generated (see Section III,A,2; Fig. 2). In any given process, the isomer produced is dependent upon the particular organism employed. This explains the special interest in the utilization of B. polymyxa for production of pure levo-diol due to its potential as an antifreeze agent (see Sections I and VIII). For a summary of the stereoisomeric nature of the diol produced by the key microbial species, refer to Table II. The possible mechanisms by which these stereoisomers are formed are discussed in Section III,A,2.

A. Klebsiella pneumoniae (Aerobacter aerogenes)

This bacterium, widely distributed in nature, is stable under a wide range of environmental conditions. Whereas K. pneumoniae lacks the diastatic properties of B. polymyxa, this species typically produces double the amount of diol, with much lower yields of by-product ethanol (Ledingham and Neish, 1954). The organism also conducts a more complete fermentation, thereby facilitating product recovery. Stock cultures of valuable strains can be stored with relative ease. For the above reasons, by far the greatest research effort in diol production has focused on the utilization of this microorganism. In addition, the capacity to produce 2,3-butanediol appears to be widely distributed within the species, so that it is conceivable that strains could be isolated with additional specific advantages, for example, the ability to hydrolyze a particular polysaccharide substrate, or the generation of a by-product animal feed with enhanced protein or riboflavin content (Bu'Lock, 1975).

B. Bacillus (Aerobacillus) polymyxa

Bacillus polymyxa has the ability to ferment a wide range of substrates. Its amylolytic nature enables its use in the conversion of

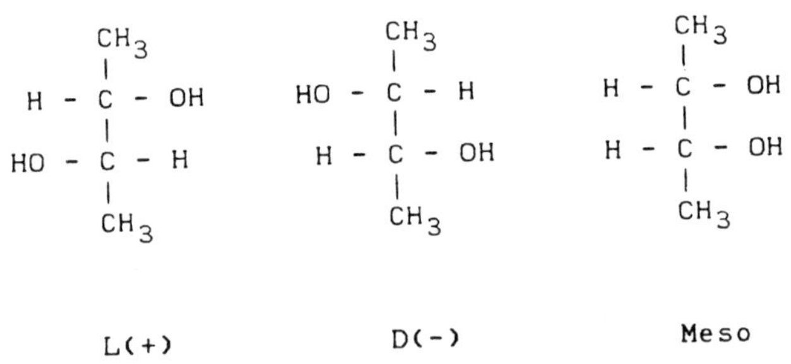

$$
\begin{array}{ccc}
\text{CH}_3 & \text{CH}_3 & \text{CH}_3 \\
| & | & | \\
\text{H} - \text{C} - \text{OH} & \text{HO} - \text{C} - \text{H} & \text{H} - \text{C} - \text{OH} \\
| & | & | \\
\text{HO} - \text{C} - \text{H} & \text{H} - \text{C} - \text{OH} & \text{H} - \text{C} - \text{OH} \\
| & | & | \\
\text{CH}_3 & \text{CH}_3 & \text{CH}_3 \\
\\
L(+) & D(-) & \text{Meso}
\end{array}
$$

FIG. 2. Stereoisomers of 2,3-butanediol.

TABLE II

BUTANEDIAOL STEREOISOMERS PRODUCED BY REPRESENTATIVE MICROORGANISMS[a]

Organism	Stereoisomer	Reference
Bacillus polymyxa	D-(−) (levo)	Neish (1945)
Bacillus subtilis	Approximately 65% D-(−), remainder meso	Blackwood et al. (1947)
Klebsiella pneumoniae	5–14% L-(+), remainder meso	Morell et al. (1944); Freeman (1947)
Serratia marcescens, Serratia spp.	Primarily meso	Neish et al. (1947, 1948)
Aeromonas (Pseudomonas) hydrophila	50% racemic, 48% meso, 2% levo	Murphy et al. (1951)
Distillers yeast	67% levo, remainder meso (possibly some dextro)	Neish (1950)

[a]Source: Ledingham and Neish (1954).

unhydrolyzed grain mashes (Ledingham et al., 1945; Fratkin and Adams, 1946). Possessing xylanase activity (Zemek et al., 1981), B. polymyxa has the potential to more fully utilize the hemicellulosic components of natural substrates (Laube et al., 1984a,b). The ability to ferment cellulose is, however, weak or lacking. Vigorous growth is seen under anaerobic conditions in the presence of fermentable carbohydrates. Interest in production of the pure levo isomer of 2,3-butanediol by this organism for possible commercial manufacture of antifreeze has already been mentioned. The formation of large quantities of by-product ethanol is generally considered to be a disadvantage. Additional problems include loss of fermentation activity due to storage or repeated transfer (Blackwood et al., 1949; Ledingham and Stanier, 1944) and a susceptibility to attack by bacteriophage (Katznelson, 1944a,b). These difficulties did not, however, prevent the utilization of this organism for the pilot-scale production of butanediol from agricultural residues (Tomkins et al., 1948; Wheat et al., 1948).

C. Bacillus subtilis

Bacillus subtilis is a second species of the genus Bacillus that has been shown to produce butanediol as a major product of the metabolism of carbohydrates. Unlike B. polymyxa, however, B. subtilis does not produce pure levo diol (Table II). Under anaerobic conditions, growth and fermentation rates are weak in comparison with other organisms. Vigorous growth, however, together with the formation of butanediol, acetoin, and carbon dioxide, has been observed when oxygen is supplied to the culture (Blackwood et al., 1947). Production of considerable quantities of glycerol, apparently at the expense of ethanol, is a common feature of carbohydrate consumption by this bacterium.

D. Serratia marcescens

A member of the same family (Enterobacteriaceae) as K. pneumoniae (Fig. 1), S. marcescens is also capable of generating reasonable yields of butanediol from monosaccharides. In addition to the diol, the formation of significant levels of organic acids (especially formic and lactic acids) typifies these fermentations. Under anaerobic conditions, these two acids account for 37% (molar basis) of the total solvents formed from glucose (Neish et al., 1947). This microbe can utilize both cellobiose and glycerol, but is not able to consume arabinose. The ability to ferment xylose is strain dependent. Several other members of the genus Serratia have also been identified as diol producers. Chief among these are Serratia indica, Serratia kiliensis, Serratia plymuthica, and Serratia anolium (Pedersen and Breed, 1928; Neish et al., 1948).

E. Aeromonas (Pseudomonas) hydrophila

Aeromonas hydrophilia is distinct from the other diol formers discussed to this point in that it is derived from a completely different order of bacteria (Fig. 1). The utilization of this facultative organism was investigated by Stanier and Adams in 1944, by Tsao (1978) for diol production from pentoses, and by Willetts (1984) for starch conversion. Under anaerobic conditions, the formation of ethanol by this organism is appreciable. It has been suggested that A. hydrophila be employed in a low-technology process for the conversion of waste materials to liquid fuels by virtue of the species' ability to generate diol in a reasonably efficient manner, without the need for elaborate pH or aeration control (Tsao, 1978).

F. OTHER ORGANISMS

Production of butanediol from glucose by two species of filamentous fungi, Rhizopus nigricans and Penicillum expansum, has been noted by

Fields and Richmond (1967). The conversion efficiency was extremely poor (approximately 0.003 g/g glucose available), and the diol productivity was limited to 0.0005 g/liter/hour.

An enteric bacterium isolated from decaying wood has been reported to be capable of converting xylose to diol at an efficiency of approximately 60% of the theoretical yield (Sankarnarayan et al., 1980; Veeraraghaven et al., 1980). The species of the organism has not been determined.

III. Biochemistry

A basic knowledge of the biochemistry of holocellulose conversion is essential in the study of the complex systems in which butanediol is a major product. A brief examination of the metabolic pathways associated with the breakdown of monosaccharides to common intermediates precedes a description of the reactions leading to the various end products. The nature of metabolic regulation is then discussed. This section serves as the basis on which to evaluate the influences on diol production of key environmental parameters.

A. Pathways of Carbohydrate Metabolism

In addition to the metabolism of glucose, the utilization of hemicellulosic sugars is included in the following discussion. Conversion of these monosaccharides to pyruvate introduces the routes involved in the generation of the major products associated with the mixed acid–2,3-butanediol fermentation. Horecker (1962) has written a general review of the metabolism of pentoses by bacteria. Unless otherwise specified, the pathways outlined here are those of bacterial metabolism.

1. Formation of Pyruvate

From glucose, pyruvate is formed in a relatively simple manner via the Embden–Meyerhof pathway (glycolysis). The reaction sequences leading to the generation of two molecules of pyruvate per molecule of glucose are illustrated in Fig. 3. In contrast, the production of pyruvic acid from five-carbon substrates must proceed via a combination of the pentose phosphate and Embden–Meyerhof pathways (Fig. 4). Pentoses and pentitols other than D-xylose can enter this route only after initial conversion to the common intermediate, xylulose-5-phosphate (Fig. 5). The net yield of pyruvate is thus 1.67 molecules per molecule of pentosan consumed. With the exception of D-ribose, the initial reaction in

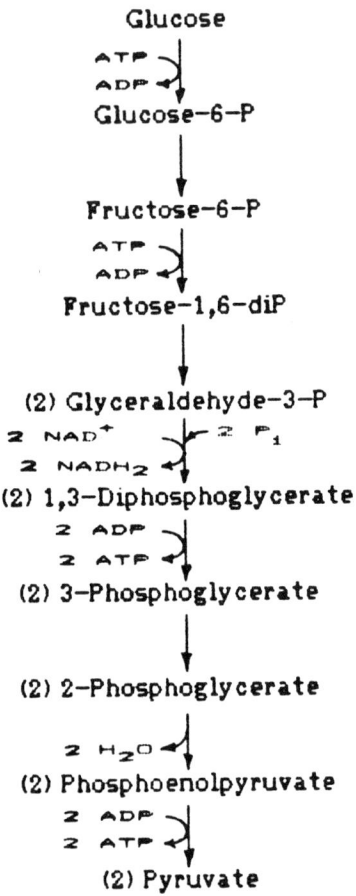

FIG. 3. Conversion of glucose to pyruvate.

the metabolism of pentoses by bacteria is catalyzed by an isomerase. This is in contrast to pentose utilization by most species of yeasts, in which oxidation–reduction of the sugar results in the formation of the corresponding pentulose (e.g., xylulose) via pentitol (e.g., xylitol). The induction in K. pneumoniae of key enzymes required for the metabolism of pentoses has been examined by Mortlock and Wood (1964).

2. *Metabolism of Pyruvate via the Mixed Acid–2,3-Butanediol Pathway*

A common feature of the bacterial conversion of carbohydrates is the

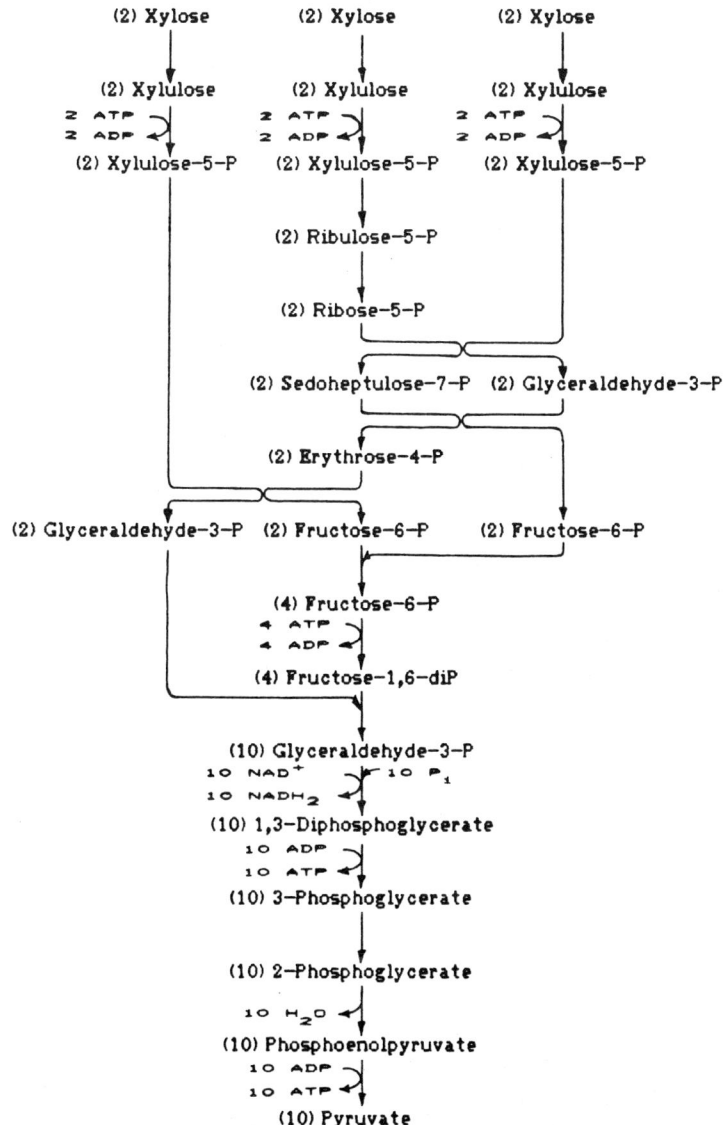

FIG. 4. Conversion of xylose to pyruvate.

formation of multiple end products. Such is the case in the production of 2,3-butanediol. In addition to the diol, ethanol is usually present (at levels that could exceed that of the diol, depending upon the bacterium

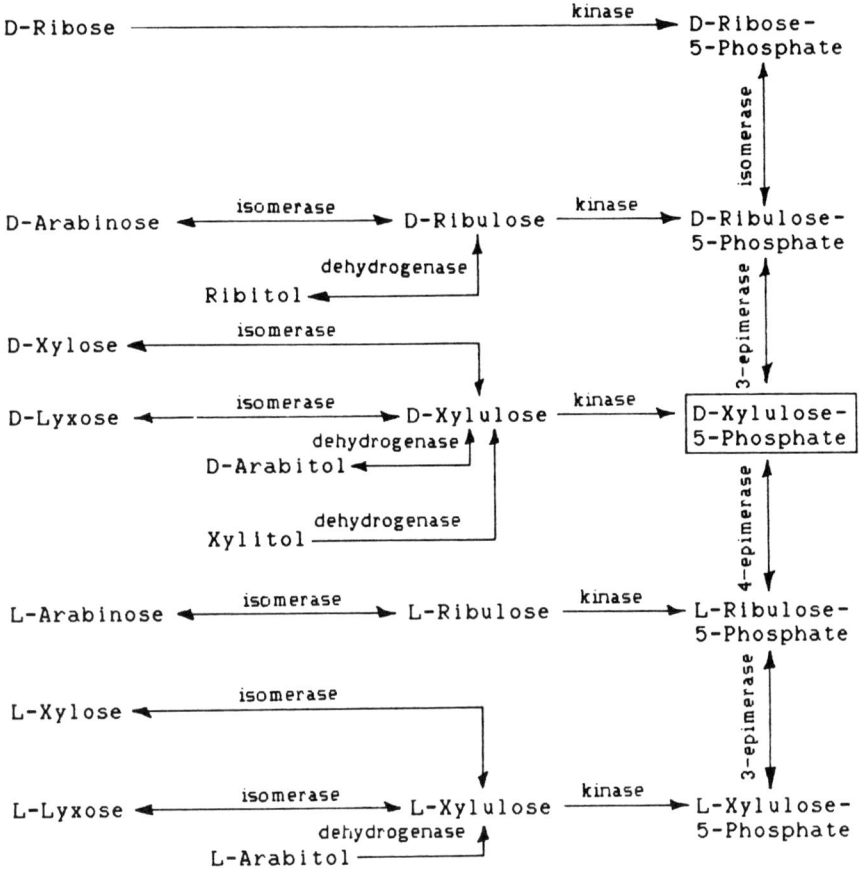

FIG. 5. Pathways of pentose and pentitol metabolism in K. pneumoniae (source, Mortlock and Wood, 1964).

employed and the culture conditions). A variety of organic acids may also be produced. The most common of these are acetate, lactate, formate, and succinate. The metabolic routes by which these diverse compounds are produced from pyruvate are illustrated in Fig. 6.

Under aerobic conditions, pyruvate is broken down to acetyl-CoA in enterobacteria by the action of the pyruvate dehydrogenase multienzyme complex. This enzyme system is not synthesized in anaerobic environments, however, and is inhibited by the reduced form of the metabolic cofactor nicotinamide adenine dinucleotide, $NADH_2$ (Gottschalk, 1979). Thus three main enzyme systems act upon pyruvate

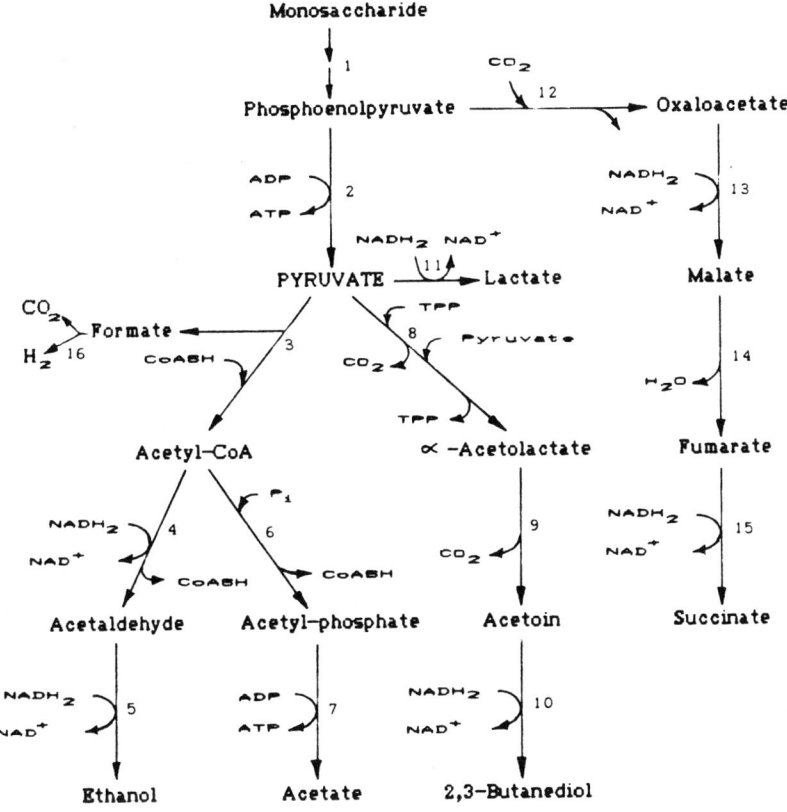

FIG. 6. Mixed acid–2,3-butanediol pathway. Enzymes: 1, enzymes of glycolysis (and pentose phosphate pathway); 2, pyruvate kinase; 3, pyruvate–formate lyase; 4, acetaldehyde dehydrogenase; 5, ethanol dehydrogenase; 6, phospho-transacetylase; 7, acetate kinase; 8, acetolactate synthase; 9, acetolactate decarboxylase; 10, acetoin reductase (butanediol dehydrogenase); 11, lactate dehydrogenase; 12, phosphoenolpyruvate decarboxylase; 13, malate dehydrogenase; 14, fumarase; 15, succinate dehydrogenase; and 16, formate-hydrogen lyase.

when the culture conditions are not fully aerobic: lactate dehydrogenase, pyruvate–formate lyase, and acetolactate synthase (pH 6 acetolactate-forming enzyme). Pyruvate–formate lyase, induced by the transition to anaerobic conditions, is relatively unstable and is inactivated when the concentration of pyruvate becomes limiting. Rapid and irreversible inactivation of this system results from the introduction of oxygen to the culture.

The formate generated by K. pneumoniae is further metabolized to carbon dioxide and hydrogen by the formate–hydrogen lyase complex. This complex is inhibited under aerobic conditions and by the presence of nitrate in an anaerobic environment (Gottschalk, 1979).

The enzymes that effect the formation of 2,3-butanediol from pyruvate are acetolactate synthase, acetolactate decarboxylase, and acetoin reductase (butanediol dehydrogenase). Acetolactate synthase is formed only under slightly acidic conditions and is thus distinct from the anabolic acetolactate synthase, or "pH 8 enzyme" (Gottschalk, 1979). The pH 6 enzyme conducts a two-step reaction. In the initial stage, pyruvate is complexed with thiamine pyrophosphate (TPP) to form acetyl-TPP. This compound is subsequently condensed with a second molecule of pyruvate to yield acetolactate. The acetolactate synthase from K. pneumoniae was purified by Stormer in 1967 and was found to have a sharp pH optimum of 5.8. Phosphate and sulfate were found to competitively inhibit the enzyme (Stormer, 1968a).

The influence of acetic acid upon the dissimilation of glucose by A. indologenes was noted by Reynolds and Werkman (1937a). These authors described a stimulation of butanediol production coupled with a decline in hydrogen evolution upon addition of the acid. Mickelson and Werkman (1939) subsequently tested homologues of acetaldehyde and acetic acid, and found that propionic acid elicited similar results. Using cell-free extracts of K. pneumoniae, Happold and Spencer (1952) similarly observed that acetic acid stimulated production of acetoin. In a further development, Stormer (1968b), working with purified preparations, showed that acetate caused induction of all of the enzymes involved in the formation of butanediol from pyruvate. Maximum activity of all three enzymes was achieved at an acetate concentration of approximately 150 mM (9.01 g/liter). In each case, activity was enhanced roughly 10-fold. When acetamide was used in place of acetate, no induction was observed. Thus it was concluded that the carboxyl group was essential for stimulation. In later trials, Stormer (1977) showed that the dissociated form of acetate (CH_3COO-) was responsible for the induction of the acetolactate synthase of K. pneumoniae. Analogs of acetate substituted at the methyl group produced similar results. It therefore appears that the functional group responsible for the inducing effect is the $COO-$ of the ionized acetate.

The optimal pH range for acetolactate synthase activity (5.5–6.5) corresponds with the pH range in which acetate is present in its dissociated form, suggesting that acetate performs an important role in the metabolic regulation of pyruvate catabolism (see Section III, B). Kinetic evaluation of the purified acetolactate synthase from S. marcescens has been studied by Malthe-Sorenssen and Stormer (1970).

Acetolactate decarboxylase isolated from K. pneumoniae has a pH optimum in phosphate buffer of 6.2–6.4 (Loken and Stormer, 1970). At a pH of 5.8 (optimal for acetolactate synthase), activity is about 75% of the maximum. The stimulatory effect of Mn^{2+} observed with crude extracts of the decarboxylase was lost upon purification of the preparation.

As has been discussed briefly in Section II, the 2,3-butanediol produced by microbial processes can exist in any of three stereoisomeric forms (the structures of which are illustrated in Fig. 2). The metabolic means by which these stereoisomers are formed has been the subject of much debate. Johansen et al. (1973) reported that a single enzyme isolated from K. pneumoniae, diacetyl (acetoin) reductase, conducted both the reversible reduction of acetoin to butanediol and the irreversible reduction of diacetyl to acetoin. This was in contrast to the work of Strecker and Harary (1954), in which a two-enzyme system was proposed: a butanediol dehydrogenase (mediating the reversible reduction of acetoin to butanediol), and a diacetyl reductase (irreversible reduction of diacetyl to acetoin).

The existence of two separate reductases was suggested by Ledingham and Neish (1954), one of which effected the reduction of D-(-)-acetoin to meso diol, while a distinct enzyme formed D-(-)-diol from D-(-)-acetoin. The possibility than an acetoin racemase existed was mentioned; however, formation of L-(+)-butanediol was not considered. A model of this mechanism is given in Fig. 7a.

Taylor and Juni (1960) proposed the existence of a three-enzyme system with separate D-(-)- and L-(+)-acetoin reductases. The reductases were said to be nonspecific with respect to acetoin stereoisomers, in that both would accept either D-(-)- or L-(+)-acetoin. The stereospecific nature of the resultant butanediol would depend upon the particular reductase involved. Thus while D-(-)-diol would be produced from D-(-)-acetoin by the D-(-)-acetoin reductase, this same enzyme would also produce meso diol when L-(+)-acetoin was used as its substrate. The corresponding activity of the L-(+)-acetoin reductase is shown in Fig. 7b.

The inability of extracts from K pneumoniae and S. marcescens to oxidize D-(-)-butanediol, and the ability of B. polymyxa and A. hydrophila to do so (Hohnbentz and Radler, 1978), indicate that acetoin reductases are in fact stereospecific. It was concluded, therefore, that K. pneumoniae and S. marcescens possessed an L-(+)-reductase, whereas B. polymyxa and A. hydrophila utilize a D-(-)-reductase. This explanation would support the data presented in Table II. A model incorporating stereospecific reductases was thus proposed by Voloch et al. (1983). This model (Fig. 7c) is suitable for organisms such as K. pneumoniae; however, extension to other species would necessitate the

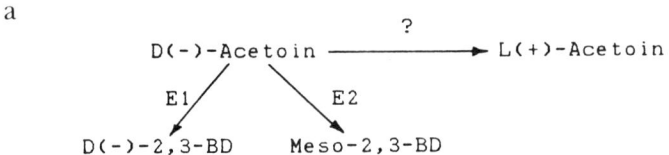

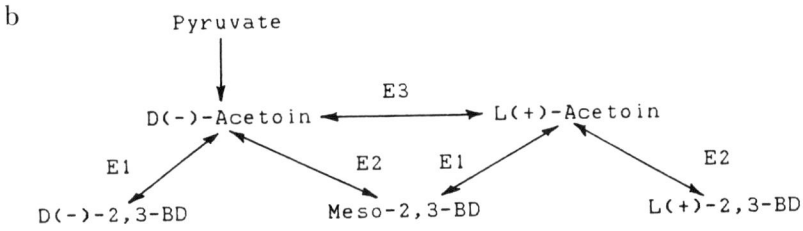

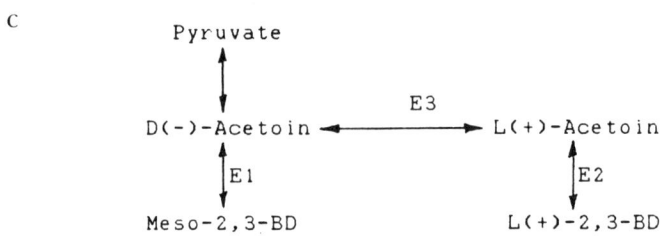

FIG. 7. Conversion of acetoin to 2,3-butanediol. (a) Model A (Ledingham and Neish, 1954); (b) Model B: E1, D-(−)-acetoin reductase; E2, L-(+)-acetoin reductase; E3, acetoin racemase (Taylor and Juni, 1960); (c) Model C; E1, D-(−)-acetoin reductase; E2, L-(+)-acetoin reductase; E3, acetoin racemase (Voloch et al., 1983).

inclusion of a second D-(−)-reductase capable of producing D-(−)-2,3-butanediol.

Low levels of acetoin reductase are found in K. pneumoniae under aerobic conditions (Johansen et al., 1975). In kinetic studies of this enzyme, Larsen and Stormer (1973) found that at pH 7.0, acetate had no effect upon the K_m for butanediol. At pH 5.8, however, the K_m was increased by a factor of 10.

The metabolic diversity of the major diol producers enables a loose classification of butanediol fermentations based on the end product whose yield is second to that of the diol (Ledingham and Neish, 1954). Thus the diol–hydrogen fermentation, in which hydrogen and carbon dioxide are generated in large quantities, is typical of such species as B. polymyxa, K. pneumoniae, A. hydrophila, and S. plymuthica (Table III).

TABLE III

CLASSES OF BUTANEDIOL FERMENTATIONS

Fermentation class	Equation	Representative organisms
Diol–hydrogen	Glucose = diol + $2CO_2$ + H_2	B. polymyxa, K. pneumoniae, A. hydrophila, S. plymuthica
Diol–formate	Glucose = diol + CO_2 + formate	S. marcescens
Diol–glycerol	2Glucose = diol + $4CO_2$ + 2glycerol	B. subtilis (Ford type)

In S. marcescens, little or no hydrogen is evolved, and the resultant accumulation of formic acid is the basis upon which this organism is categorized as a diol–formate producer. Production of significant quantities of glycerol, characteristic of Ford strains of B. subtilis, is noted in the final class. Product yields from the anaerobic dissimilation of glucose are presented in Table IV for representative organisms of these three groups.

B. BIOCHEMICAL ROLE OF THE 2,3-BUTANEDIOL PATHWAY

Two important metabolic functions have been attributed to the formation of butanediol in microbial species. Regulation of intracellular pH was proposed by Stormer (1968b). Prevention of acetate overproduction and subsequent acidification of the intracellular pH is mediated via the sensitivity of the acetolactate synthase to induction by dissociated acetate. Thus pyruvate is channeled into diol production when acetate reaches a critical concentration.

Maintenance of the intracellular $NAD^+/NADH_2$ balance has been suggested by Johansen et al. (1975) to be an important role of diol formation. Reduction of acetoin to butanediol serves to regenerate NAD^+ from $NADH_2$ during the active catabolism of carbohydrates. When exhaustion of the substrate occurs, $NADH_2$ is no longer formed, and the reversibility of the acetoin reductase reaction permits replenishment of $NADH_2$ for cellular function. Butanediol thus serves as a reservoir of reducing equivalents.

IV. Environmental Parameters

Environmental factors influencing the bacterial production of 2,3-butanediol are discussed briefly in this section. Critical in all

TABLE IV

PRODUCT BALANCES IN ANAEROBIC DISSIMILATION OF GLUCOSE[a]

	Organism			
Product	Bacillus polymyxa	Klebsiella pneumoniae	Serratia marcescens	Bacillus subtilis (Ford type)
Butanediol	32.6	24.0	32.0	27.5
Hydrogen	0.8[b]	0.4[b]	—	—
Formic acid	—	0.8	12.3[b]	0.3
Glycerol	—	—	—	29.4[b]
Ethanol	17.1	17.2	11.8	2.0
Lactic acid	—	2.4	5.0	11.6
CO_2	48.2	43.8	29.6	26.9
Total	98.7	88.7	90.7	97.7

[a]As g/100 g glucose consumed. Source: Ledingham and Neish (1954).
[b]Identifies class of fermentation based on end product.

fermentations, such parameters as pH, temperature, and product and substrate concentration are evaluated in addition to those elements that have proven to be of particular importance to processes in which diol is the principal end product (aeration, acetate supplementation, and water activity).

A. CARBON SOURCE

Whereas the initial work on diol production focused upon the nature of glucose dissimilation, the need was soon recognized for the evaluation of all those sugars present in feedstocks judged to be of commercial potential. Thus the principal monosaccharides of cellulosic substrates (glucose, mannose, and galactose) were examined, as were the major hemicellulosic components (xylose and arabinose). Early studies of xylose utilization (Reynolds and Werkman, 1937), were instrumental in identifying the importance of aeration in the efficient production of butanediol. The potential for diol manufacture from sugar crops (molasses, sugar beets, etc.) stimulated interest in the bioconversion of sucrose (Pirt and Callow, 1958, 1959).

The relative efficiencies of diol production from each of these monosaccharides are briefly discussed here. A summary of solvent yields and productivities typical of diol formation from these carbohydrates is provided in Table V. Note that the results presented are for batch culture under anaerobic conditions only. A discussion of the influence of aeration and process design upon diol yields can be found in Sections IV,G and VI, respectively. For an examination of the conversion of potential commercial feedstocks, refer to Section V.

Under anaerobic conditions, typical yields of butanediol from the batch dissimilation of glucose are in the range 0.35 g/g of carbohydrate available (approximately 70% of the theoretical yield). Ethanol production, on a molar basis, is nearly equal to that of the diol. Organic acid synthesis is usually limited. In contrast, yields of acetic acid from xylose frequently exceed diol production in the absence of oxygen. This is in part, however, a reflection of the extremely poor efficiency of diol formation (less than 0.10 g/g xylose available). Ethanol formation under these environmental conditions far exceeds that of diol, the molar ratio of ethanol to diol in the culture broth being approximately 5 to 1. Near-equimolar yields of ethanol and diol are observed from arabinose, galactose, and mannose in anaerobic batch culture. Diol yields are approximately 50% of the theoretical yield for each of these monosaccharides.

Yu and Saddler (1982a) tested the carbohydrates listed in Table V for diol production under aerobic, anaerobic, and finite air conditions (small amount of air initially available in sealed flask, rapidly exhausted as growth occurs). Optimal production of diol was observed in the absence of oxygen for each of mannose, galactose, arabinose, and cellobiose. The influence of aeration on the dissimilation of glucose is examined in Section IV,G. Biomass formation under anaerobic conditions was lowest when xylose was used as substrate, indicative of the correlation between butanediol and biomass yields.

B. Temperature

Optimization of process temperature is recognized to be an essential aspect of successful design. The strong dependence of enzymatic activity and cellular maintenance requirements upon temperature (Esener et al., 1983) makes the effeciency of bioprocesses strictly temperature dependent.

The optimum temperature for the bacterial formation of 2,3-butanediol is generally reported to be in the range 30–35°C. Blanco et al. (1984) recorded optimum diol production from glucose at 34°C. Using sucrose as the carbon source, Pirt and Callow (1958) observed that diol production was maximal between 35 and 37°C. The optimum temperature for sucrose uptake by K. pneumoniae was similarly found to be 35°C.

TABLE V

ANAEROBIC DISSIMILATION OF PURE CARBOHYDRATES[a]

Substrate		Temp. (°C)	Solvent yields (g/liter)				Diol (+ AMC)		Ref.[b]
Sugar	g/liter		Diol	AMC	EtOH	HAc	(g/g CHO available)	Productivity (g/liter/hr)	
Glucose	20	30	7.4	0.1	2.7	0.2	0.37	0.08	1
Glucose	10	37	3.5	—	1.8	0.1	0.35	0.14	2
Xylose	10	37	0.4	—	1.0	0.9	0.04	0.02	2
Arabinose	10	37	2.5	—	1.4	0.3	0.25	0.11	2
Galactose	10	37	2.3	—	1.4	0.3	0.23	0.09	2
Mannose	10	37	2.7	—	1.5	0.3	0.27	0.11	2
Cellobiose	10	37	1.1	—	1.3	0.5	0.11	0.05	2

[a] Batch culture; *K. pneumoniae*.
[b] References: (1), Mickelson and Werkman (1938) and (2), Yu and Saddler (1982a).

TABLE VI

INFLUENCE OF pH ON THE DISSIMILATION OF 2% GLUCOSE[a]

Product	Yield (g/liter)	
	pH maintained above 6.3 (~7)	pH maintained at or below 6.3
Butanediol	2.02	7.39
Acetoin	0.00	0.09
Ethanol	3.45	2.69
Acetic acid	6.55	0.20
Formic acid	4.37	0.56
Latic acid	0.53	0.22
Total	16.92	11.15
AMC + diol	2.02 (12%)	7.48 (67%)
Organic acids	11.45 (68%)	0.98 (9%)

[a]Culture conditions: 4-liter Erlenmeyer flask; 30°C; anaerobic; manual pH control. Source: Mickelson and Werkman (1938).

Production of butanediol is generally accepted to be a growth-associated phenomenon. Thus conditions for optimal product formation should approximate those of maximum biomass yield. It is not surprising, therefore, that Esener et al. (1981a), using an Arrhenius-type model, predicted that the maximum specific growth rate for K. pneumoniae should occur at 37°C. An unexpected result, reported by Laube et al. (1984a), was that diol production by B. polymyxa from 1% xylose was constant at 25, 30, and 35°C.

C. pH

Like temperature, pH is a fundamental parameter in the regulation of bacterial metabolism. Its influence is especially pronounced in processes characterized by the formation of multiple end products. As a general rule, alkaline conditions favor the formation of organic acids, with a corresponding decline in the yield of such products as butanediol.

This trend is apparent in studies of the effect of pH upon the anaerobic dissimilation of 2% glucose (Mickelson and Werkman, 1938). Product spectra obtained when (1) the pH was maintained near 7 by the manual addition of 1 N NaOH and (2) the pH was kept at or below 6.3 are presented in Table VI. The predominance of organic acid production in the first case is clearly evident. Yields of butanediol and acid (expressed as the sum of acetate, formate, and lactate production) are 12 and 68%, respectively, of the total solvents formed (weight basis). In contrast, under acidic culture conditions, organic acid synthesis is reduced more than 10-fold, whereas diol production is improved by a factor of 3.7. The net result is a reversal of relative yields, with diol accounting for 67% of the total solvent production and acids limited to 9%.

Production of acetic acid from xylose is at a minimum in the pH range 5.2–5.6 and increases rapidly when the pH exceeds 6 (Jansen, 1982). A 10-fold increase in the yield of lactate from glucose is similarly observed when the pH is increased from 5.6 to 7.0 (Neish et al., 1948).

The pH optimum for diol production appears to be a function of the particular substrate employed. With glucose, the optimum pH is 6.4 (Blanco et al., 1984) in contrast, when xylose is used as the carbon source, diol production is greatest at a pH of 5.0–5.2 (Tsao, 1978). The establishment of conditions for the efficient bioconversion of complex feedstocks is clearly complicated by this observation.

The connection between biomass and diol yields is also evident from pH data. Under aerobic conditions, the maximum specific growth rate (0.61/hour) was observed to occur at pH 5.2 when 10% xylose was the substrate (Jansen et al., 1984). While maximal diol formation was also noted at this pH, the yields and distribution of products were not found to be significantly affected in the pH range 4.8–5.6. No growth was observed when the pH was below 4.2.

D. Substrate Concentration

Under finite air conditions, significant inhibition of both diol formation and sugar utilization has been observed at carbohydrate levels exceeding 50 g/liter (Yu and Saddler, 1983). While this trend was particularly evident when D-xylose was the carbon source, the effect was also noted in studies utilizing D-glucose, L-arabinose, D-galactose, D-mannose, and D-cellobiose.

Improvements in diol yield were, in contrast, observed with increasing glucose concentration when K. pneumoniae was grown in an aerobic environment (Sablayrolles and Goma, 1984). At an initial substrate level of 195 g/liter, diol formation reached 90% of the theoretical yield at a productivity approximately equal to that recorded in the presence of 5% initial glucose (Table VII).

Aside from the aeration conditions employed, the major difference between these two studies lies in the amount of nitrogen available in their respective media. Both investigations employed chemically

TABLE VII
Effect of Glucose Concentration on Product Yields[a]

Initial glucose (g/liter)	Solvents (g/liter)		Butanediol		Biomass		μ max (hr^{-1})
	Diol	EtOH	g/g	g/liter/hr	g/liter	g/g	
22	7.2	0.54	0.32	0.76	4.60	0.21	0.66
45	14.5	1.72	0.32	1.16	6.95	0.15	0.67
107	44.5	2.00	0.42	2.02	8.60	0.08	0.48
195	88.0	3.23	0.45	1.10	8.50	0.045	0.19

[a]Source: Sablayrolles and Goma (1984).

defined media in which ammonium sulfate was used as a nitrogen source. Sablayrolles and Goma also included ammonium phosphate in their nutrient formulation, so that the final concentration of ammonium, at 0.200 mol/liter, is more than five times the amount found in Yu and Saddler's medium. In controlled studies under finite air conditions, Yu and Saddler (1982b) demonstrated that inhibition of diol yields from 4% glucose was the net result of a doubling of the ammonium sulfate content of their medium (Table IX). Thus it appears that the main factor in the efficient conversion of concentrated glucose solutions is the adequate availability of oxygen.

The initial glucose level appears to affect both biomass and product formation. The biomass yield (gram/gram) at 195 g/liter initial glucose is less than one-quarter that observed from 22 g/liter glucose (Table VII). The maximum specific growth rate of the culture (μ_{max}) can also be seen to be severely depressed by high glucose concentrations. Whereas diol yields have been shown to be enhanced under aerobic conditions at high glucose concentrations, the production of ethanol is depressed under these environmental circumstances. At 195 g/liter glucose, the ratio of diol to ethanol is more than double that at 22 g/liter glucose. The maximum specific rate of diol production (ν_{max}) is, however, reduced at high glucose levels.

Seemingly conflicting reports exist concerning the influence of initial xylose concentrations upon diol yields. Laube et al. (1984a) indicated a steady decline in diol yields (gram/gram) as xylose levels were increased to 60 g/liter. At initial xylose concentrations of 50, 100, and 150 g/liter, virtually no change was observed by Yu and Saddler (1983) in the yield of diol per gram of carbohydrate available. Jansen et al. (1984) reported that diol yields improved as a result of increases in the xylose

TABLE VIII
EFFECT OF XYLOSE CONCENTRATION ON PRODUCT YIELDS

Initial xylose (g/liter)	Solvent yields (g/liter)			Diol (+ AMC)		Reference
	Diol	AMC	EtOH	g/g	g/liter/hr	
20	4.0	1.8	0.9	0.29	0.081	Laube et al. (1984a)
40	6.6	3.6	1.0	0.26	0.085	Laube et al. (1984a)
60	6.4	5.4	0.7	0.20	0.082	Laube et al. (1984a)
100	0.8	NR[a]	NR	—	—	Laube et al. (1984a)
5	0.0	NR	NR	0.00	0.00	Jansen et al. (1984)
10	0.5	NR	NR	0.05	0.06	Jansen et al. (1984)
20	3.0	NR	NR	0.14	0.38	Jansen et al. (1984)
50	11.0	NR	NR	0.22	1.00	Jansen et al. (1984)
100	29.6	NR	NR	0.30	1.35	Jansen et al. (1984)
150	49.0	NR	NR	0.33	1.09	Jansen et al. (1984)
50	13.9	3.5	2.5	0.35	—	Yu and Saddler (1983)
100	23.8	9.3	2.4	0.33	—	Yu and Saddler (1983)
150	44.3	6.1	3.3	0.34	—	Yu and Saddler (1983)

[a]NR, Not reported.

concentration from 5 to 150 g/liter. These results are summarized in Table VIII.

The major factor contributing to the disparity between the results of Laube et al. and Jansen et al. is likely the choice of organism employed in the two studies. While Jansen et al. (1984) (and Yu and Saddler) used K. pneumoniae in their work, Laube et al. investigated the effects of substrate concentration on the dissimilation of xylose by B. polymyxa. The influence of xylose concentration thus appears to be species dependent.

The source of the conflicting observations of Jansen's and Yu's groups must lie in the culture conditions of the two studies. While both investigations employed chemically defined media, acetic acid (0.5%) was included in that of Yu and Saddler. Their medium's pH was initially set at 6.5 and was buffered in the approximate pH range of 5.5–6.5. Shake flasks at 30° were utilized by this group. In the studies of Jansen et al. (1984), the culture pH was automatically controlled at 5.2 in the presence of intense aeration (1.1 VVM, where VVM is volume of air/unit volume of culture/minute) in a 7-liter bioreactor at 37°C. This aeration rate appears to have been in excess of the optimal rates for low substrate concentrations (i.e., less than 50 g/liter). It is only at these low xylose levels that the trends observed by the two groups disagree significantly. At initial xylose levels in excess of 50 g/liter, both studies indicate diol yields between 0.30 and 0.35 g/g.

These findings accentuate the importance of aeration in the conversion of xylose, in particular, to butanediol. Jansen et al. (1984) noted that the primary influence of the initial xylose concentration (under the conditions of their studies) appeared to be exerted through an effect on cell yields. Elevated biomass densities as a result of the high substrate levels produced a decline in the specific oxygen supply rate, and thus an improvement in the diol yield. The maximum specific growth rate observed by Jansen et al. (1984) (1.05/hour) was recorded at a xylose concentration of 20 g/liter. No growth was detected when the xylose level exceeded 160 g/liter.

E. Product Concentration

There appears to be general agreement in the literature that the major influence of high butanediol concentrations is on biomass formation and not on product yields. Sablayrolles and Goma (1984) observed that the maximum specific growth rate of K. pneumoniae reached an upper limit at a low diol level (less than 1 g/liter) and decreased rapidly when this level was exceeded. The specific rate of butanediol formation was, in contrast, hardly affected by diol concentrations up to 80 g/liter. In shake flask cultures with an 8% initial xylose, Jansen et al. (1984) noted that no growth occurred when the diol concentration exceeded 65 g/liter. The effect of the diol on product formation by the culture was not reported. The influence of butanediol upon B. polymyxa appears to be similar in nature to that on K. pneumoniae. Laube et al. (1984a) observed that initial diol levels of up to 20 g/liter had no effect upon either the yield of diol or the consumption of the 1% xylose substrate by the B. polymyxa strain examined. It appears, therefore, that butanediol, at high concentrations, strongly inhibits bacterial growth but has very little effect upon the metabolic pathways leading to its formation.

Whereas butanediol does not exert strong feedback control of its own formation, this does not appear to be similarly true of all products of the diol pathway. Ethanol, added at the 1% level to K. pneumoniae cultures growing on 2% glucose, caused a 45% reduction in the yield of diol after 24 hours of growth. This inhibition was, however, completely overcome with time (Yu and Saddler, 1982b). Similar effects were observed upon the addition of 1% acetic acid. Although the initial inhibition due to the acetate was significantly greater (75%), this too was completely overcome. Addition of lower concentrations of acetate was found to enhance diol production. The mechanism and extent of this stimulatory effect are discussed in Sections III,A,2 and IV,F,3.

F. Medium Supplements

Culture media must contain all the essential nutrients for the growth and maintenance requirements of the particular organisms for which they are formulated. In addition to sources of carbon and nitrogen, a minimal medium may include vitamins, trace metals, etc.; the selection of which may in part be determined by the nature of the desired fermentation products. Such is the case in the formation of butanediol, where certain metallic cations have been found to improve conversion efficiencies. The influence of several supplements on yields of diol will be examined here.

1. Yeast Extract

The synthesis of protein is a constant function of cellular metabolism, both for the generation of new biomass and for the replacement of enzymes subject to rapid turnover. As a major component of protein, nitrogen must be supplied in large quantities, thus the inclusion in media of rich sources of nitrogenous compounds such as urea, peptone, and yeast extract.

In the dissimilation of 1% glucose by *B. polymyxa*, Laube et al. (1984a) found that increasing the concentration of yeast extract (YE) from 0.3 to 1.2% (w/v) had no effect on the yield of butanediol. At higher substrate levels, however, addition of YE was observed to be beneficial. Optimal diol production from 5% glucose was found to occur in the presence of 1.5% YE (Laube et al. 1984b). The diol yield observed was 18 g/liter after 3 days, versus 5.4 g/liter in the presence of 0.5% YE. Yu and Saddler (1982b) reported similar gains in diol yield from 4% glucose with *K. pneumoniae*, as a result of YE enrichment of defined medium. Production of diol, ethanol, and acetate were all enhanced by the inclusion of YE at the 1% level. After 2 days of incubation under finite air conditions, reported yields of these solvents were 12.40, 2.35, and 0.64 g/liter, respectively. Ethanol and diol production were improved by 26 and 51%, respectively, whereas acetate yields were more than tripled.

Even at low substrate concentrations, xylose bioconversion by *B. polymyxa* is facilitated by the addition of YE to the culture medium, indicative of the higher nutritional requirements of pentose utilization. Diol yields in the presence of 1.2 and 0.3% YE were 2.2 and 0.5 g/liter, respectively, after 2 days, from 1% initial xylose (Laube et al., 1984a). This enhanced production is, however, only 44% of the theoretical yield. A 50% improvement in diol yield from 4% xylose accompanied the addition of 1% YE to cultures of K. pneumoniae (Yu and Saddler, 1982b). Ethanol production was also stimulated (41% increase), whereas acetate levels declined by 29%. The yields of these three solvents were thus 11.30, 2.59, and 0.42 g/liter after 2 days.

Although yeast extract was found to enhance diol yields from both glucose and xylose, its high cost prohibits the utilization of large quantities in commercial processes. Attempts have been made, therefore, to displace part of the extract with other media supplements. Laube et al. (1984b) examined the ability of various nitrogen sources, vitamins, and trace metals to substitute for two-thirds of the YE level (1.5%) found to be optimum for 5% glucose consumption. Only the trace metals (see below) proved effective in this regard. The nitrogen sources tested included urea (1 and 0.5%), ammonium sulfate (0.1 and 0.3%), and cysteine monohydrate (0.2%).

Yu and Saddler (1982b) found that the addition of 1% urea to a defined medium resulted in a 28% improvement in the yield of diol from 4% glucose (10.49 g/liter after 2 days). This increase was less, however, than that produced by the supplementation of the medium with 1% YE. Improvements in xylose conversion by the presence of urea were of similar magnitude. Doubling the ammonium sulfate level of the glucose medium (from 3.0 to 6.0 g/liter) resulted in a 16% reduction in diol yield. Ethanol yields were similarly reduced, whereas acetate formation was tripled. When the ammonium sulfate content was doubled in medium containing 4% xylose, the effect on conversion efficiencies was opposite to that of the glucose. Diol yields were improved by 45%, resulting in 10.91 g/liter diol (versus 11.30 g/liter in the presence of 1% YE). The production of acetic acid was reduced by 25% and ethanol yields were slightly enhanced (7%). These observations are summarized in Table IX.

In the conversion of glucose by K. pneumoniae, B. subtilis, and B. polymyxa, Mahmoud et al. (1975) found proteose–peptone to be superior to yeast extract as a source of nitrogen. Diol yields with K. pneumoniae grown in the presence of proteose–peptone, yeast extract, and urea were 16.1, 14.6, and 15.1 g/liter, respectively, from 5% glucose. The final nitrogen content was the same in each of the media preparations. The optimum nitrogen concentration for diol production was found to be 0.07%. Higher levels, although not inhibitory to diol formation, did not further enhance the product yields.

TABLE IX

EFFECT OF NUTRIENT SUPPLEMENTS ON SOLVENT YIELDS AND PRODUCTIVITIES FROM GLUCOSE AND XYLOSE

Substrate		Supplement[a]	Aeration[b]	Solvent yields (g/liter)			Diol yield		Diol Productivity (g/liter/hr)	Ref[c]
Sugar	g/liter			Diol	EtOH	HAc	g/g	%		
Glucose	40	Control	Fin. air	8.2	1.9	0.2	0.21	41	0.17	2
Glucose	10	Control	Anaerobic	3.5	1.8	0.1	0.35	70	0.14	1
Glucose	40	1% YE	Fin. air	12.4	2.4	0.6	0.31	62	0.26	2
Glucose	40	1% urea	Fin. air.	10.5	2.5	0.8	0.26	53	0.22	2
Glucose	40	2(NH$_4$)	Fin. air	6.9	1.6	0.6	0.17	35	0.14	2
Glucose	40	10(tr. m)	Fin. air	9.6	1.8	0.5	0.24	48	0.20	2
Glucose	40	0.5% HAc	Fin. air	19.6	NR[d]	NR	0.49	83[e]	0.82	2
Xylose	40	Control	Fin. air	7.6	1.8	0.6	0.19	38	0.16	2
Xylose	40	1% YE	Fin. air	11.3	2.6	0.4	0.28	56	0.24	2
Xylose	40	1% urea	Fin. air	9.8	2.4	0.4	0.24	49	0.20	2
Xylose	40	2(NH$_4$)	Fin. air	10.9	2.0	0.4	0.27	55	0.23	2
Xylose	40	10(tr. m)	Fin. air	8.4	1.8	0.6	0.21	42	0.17	2
Xylose	40	0.5% HAc	Fin. air	22.0	4.2	1.0	0.55	93[e]	0.46	2

[a]2(NH$_4$), double ammonium sulfate content of control medium; 10(tr. m), 10 times trace metal content of control medium.
[b]Fin. air, Finite Air.
[c]References: (1), Yu and Saddler (1982a) and (2), Yu and Saddler (1982b).
[d]NR, Not reported.
[e]Theoretical yield based on carbohydrate + initial acetate.

2. Trace Metals

Yeast extract contains the cationic species of the following metals: Al, Ba, Cd, Cu, Cr, Fe, Mg, Mn, Mo, Ni, Pb, Ti, V, and Zn (Grant and Pramer, 1962). Supplementation of a medium containing 0.5% YE to a final trace metals content equivalent to that of 1.5% extract resulted in improved diol yields (Laube et al., 1984b). The reported yields of butanediol from 5% glucose in medium containing 0.5% YE, 0.5% YE + trace metals, and 1.5% YE were 6.7, 14.6, and 17.5 g/liter, respectively.

The principal stimulatory components of the trace metals mixture were found to be Fe^{2+} and Mn^{2+}. Enhancement of diol production by the additon of Fe had been reported previously by Roberts (1947), and Mn^{2+}-induced increases were noted by Sankarnarayn et al. (1980). The concentrations of these cations found by Laube et al. (1984b) to give maximal stimulation were 40 and 1.7 μM, respectively. Supplementation of medium containing 0.5% YE (control yield = 6.6 g/liter) with Fe, Mn, and Fe^{2+} plus Mn^{2+} resulted in yields of 8.7, 12.6, and 13.0 g/liter, respectively. The yield in the presence of 1.5% YE (17.8 g/liter) was not, however, achieved by the supplementation of the medium with these factors. The Mn^{2+}-induced enhancement of diol formation was initially believed to result from the induction of acetolactate decarboxylase (Sokatch, 1969). Recall, however, that Loken and Stormer (1970) found that Mn^{2+} did not stimulate a highly purified preparation of this enzyme from K. pneumoniae.

Phosphate has been shown by several authors to stimulate diol production (Murphy et al., 1981a; Taha et al., 1971). The addition of potassium phosphate (or sodium phosphate) to media containing 0.5% YE was found to improve conversion efficiencies (Laube et al., 1984b). The optimum concentration of the phosphate was 78 μM, resulting in an average diol yield of 13.8 g/liter from 5% glucose. Through combination of these stimulatory factors (Fe^{2+}, Mn^{2+}, and phosphate), the resultant yield of butanediol (15.4 g/liter) was 89% of that obtained in the presence of 1.5% YE.

3. Acetate

In Section III,A,2, the induction by acetate of the enzymes which catalyze the breakdown of pyruvate to butanediol was discussed briefly. The effects of various concentrations of acetate upon the product and biomass yields in the bioconversion of glucose by K. pneumoniae are described in Section IV,E. Thus, acetic acid is of special interest in the production of butanediol, as it is both a metabolic by-product of carbohydrate consumption and an important medium supplement for the enhancement diol yields.

The marked improvement in the efficiency of diol formation from both glucose and xylose under finite air conditions can be seen from Table IX. Taking into account the potential for diol formation from the acetate itself (via condensation with pyruvate), the yield of diol from 4% glucose was more than doubled by the inclusion of 0.5% acetate in the medium. The production of diol thus reached 83% of the theoretical yield. The effects of acetate supplementation were greater still when xylose was the carbon source. In the absence of any nutrient supplementation, the yield of diol from 4% xylose was limited to 38% of the theoretical value. Incorporation of 0.5% acetate into the medium resulted in the accumulation of diol at 93% of the theoretical yield. The productivity of diol from xylose (grams/liter/hour) was nearly tripled, whereas from glucose the rate of diol formation was increased almost fivefold.

The enhancement of diol yield and productivity by the presence of acetate appears to be species dependent. Laube *et al.* (1984a), studying the bioconversion of 1% xylose by *B. polymyxa*, observed no effect due to the initial incorporation of 0.5 or 1.0% (w/v) acetate into the culture medium.

G. Aeration

The primary microbial species known to produce significant quantities of 2,3-butanediol from carbohydrates (*K. pneumoniae, A. hydrophila, B. subtilis,* and *B. polymyxa*) are classified as facultative anaerobes and, as such, can exist in the absence of oxygen. Diol production has been shown, however, to be most efficient under aerobic conditions (especially when high substrate concentrations are encountered, or during the "fermentation" of five-carbon sugars).

Regeneration of ATP and NAD^+ is a major function of molecular oxygen in the growth and maintenance of aerobic and facultative organisms. Oxygen is the terminal electron acceptor in the production of ATP via oxidative phosphorylation. This compound subsequently provides the energy required in many of the reactions of cell synthesis, maintenance, and product formation. In the initial reactions of the electron transport chain, by which electrons are passed to oxygen with coupled ATP formation, NAD^+ is regenerated from $NADH_2$. The NAD^+ is an essential cofactor of the many dehydrogenases involved in the oxidation–reduction reactions of the catabolic and biosynthetic pathways. Both ATP and NAD, in addition, act as key regulators in the control of bacterial metabolism (Sanwall, 1970). The formation of compounds such as butanediol is thus mediated by the presence of oxygen in the culture environment.

Methods of quantifying microbial system oxygenation are as varied as the apparati employed in the cultivation procedures. With simple vessels such as Erlenmeyer flasks fitted with porous plugs, the primary parameters affecting the degree of aeration are the surface area to volume ratio of the culture and the speed at which the flasks are shaken. In cases in which the culture within such a vessel is sparged with air, the flow rate (usually expressed as the volume of air/unit volume of culture/minute; VVM) can be controlled as well. In contrast, more elaborate bioreactors regulate aeration primarily through the design and operational characteristics of the mixing apparatus. Although the geometry of the reactor is of considerable importance, the design and placement of impellers and baffles within the vessel are the primary parameters influencing the transfer of air to the medium. Although impeller speed is occasionally cited as an indicator of culture aeration, the dissolved oxygen content of the medium as measured by sterilizable electrodes is more frequently reported. This level may be expressed either as a dissolved oxygen tension (DOT) with units of pressure (millimeters Hg), or as a percentage of the saturation level of the particular medium employed. Alternatively, the rate of oxygen transfer to the culture, Q_{O_2} (millimoles oxygen/liter/hour), or the oxygen absorption coefficient, $K_L a$ (hr^{-1}), may be reported. Such measurements are not usually made directly on the culture, but on a synthetic solution resembling the culture as closely as possible under identical conditions of aeration, agitation, temperature, etc. To be truly representitive, the solution employed must have the same viscosity, rheology, bubble coalescence, gas–liquid interfacial resistance, oxygen solubility, and diffusivity as the actual medium (Bailey and Ollis, 1977). Theoretically, the oxygen uptake (absorption) rate of the solution (Q_{O_2}) is equal to the product of the $K_L a$ and the difference between the saturation level of the solution (C^*) and the actual concentration of oxygen (C_L):

$$Q_{O_2} = K_L a (C^* - C_L) \tag{1}$$

where K_L is the overall liquid film mass-transfer coefficient (m^3/m^2/sec; or m/sec), and a is the specific area of the gas–liquid interface (surface area per unit volume; (m^2/m^3; or m^{-1}). In the sulfite oxidation method of $K_L a$ determination, C_L is maintained essentially at zero through rapid consumption of oxygen in the presence of a metallic catalyst as shown in Eq. (2):

$$SO_3^- + \tfrac{1}{2} O_2 + e^- \overset{Co^{2+}}{\rightleftarrows} SO_4^{2-} \tag{2}$$

Q_{O_2} is then determined from the rate of the chemical reaction, and $K_L a$ can therefore be calculated from Eq. (1) if the saturation concentration of oxygen C^* is known. Under these conditions the rate of oxygen transfer to the solution is maximal, since the driving force, $C^* - C_L$, is greatest when C_L is zero.

In a microbial culture, the maximum rate of oxygen consumption is given by the following equation:

$$\text{Maximum Oxygen Utilization Rate} = X\mu_{max}/Y_{O_2} + Xm_{O_2} \tag{3}$$

where X is the cell density (grams/liter), μ_{max} is the maximum specific growth rate (hr^{-1}), Y_{O_2} is the cell yield per mole of oxygen consumed (grams/mole oxygen), and m_{O_2} is the maintenance respiration rate (moles oxygen/grams cells/hour).

When the dissolved oxygen content of the medium exceeds the critical oxygen concentration, the cell is saturated with oxygen and the process is said to be biochemically limited. If, however, the reverse situation is observed and the supply of oxygen to the culture does not meet the cellular demand, the system is mass-transfer limited, and the $K_L a$ must be improved if biomass or product yields are to be enhanced.

A further method of quantifying aeration occasionally reported employs polarographic or gas chromatographic analysis of the reactor exit gas composition coupled with accurate measurement of gas flow rates. By this means, it is possible to calculate the respiration rate of the culture (mmoles oxygen/grams cells/hour). In the literature dealing with the influence of aeration on diol production, all of the above systems of measurement can be encountered (surface to volume ratio; rpm; VVM; DOT; $K_L a$; oxygen supply rate; and relative respiration rate). A comparative evaluation of aeration effects is thus extremely difficult.

Adams and Leslie (1946) found that in studies of the conversion of wheat mashes by *B. polymyxa*, improvements in the diol yield could be obtained by increasing the surface area to volume ratio of shake flask cultures. Similar yield increases resulted when the experiments were repeated under a nitrogen atmosphere or under reduced pressure. The conclusion was drawn that the enhanced diol production was the result of carbon dioxide removal, and not due to a direct effect of the oxygen. Growth of K. pneumoniae under reduced pressure did not, however, improve diol production, and thus CO_2-stripping of the medium could not explain the positive effect of aeration upon this organism.

In the batch culture of K. pneumoniae on 10% glucose, Olson and Johnson (1948) found that optimal diol production occurred when a stepwise reduction of aeration rate was effected. During the first 12 hours, biomass formation was stimulated by an aeration level of 0.59

VVM. A step change to 0.39 VVM was then maintained for 24 hours, after which the aeration rate was further reduced to 0.20 VVM. This lower rate was maintained until the fermentation was complete. The final combined yield of acetoin and butanediol as a result of this treatment was 0.42 g/g (42.0 g/liter) or 84% of theory. Productivity was 0.52 g/liter/hour. Repetition of the growth under identical conditions except for the utilization of an aerated inoculum resulted in similar diol yields (40.5 g/liter), but at almost double the rate of diol formation. A markedly improved productivity of 0.92 g/liter/hour was thus observed.

For every microbe a "critical" dissolved oxygen level (C_{CRIT}) exists above which the organism's respiration rate is independent of the oxygen concentration. The value of this critical oxygen concentration is in the range 0.003–0.05 mmol/liter for most microbes (Kinn, 1967). This range is approximately 2–30% of the oxygen solubility of air-saturated nutrient solutions. For *Escherichia coli* at 37.8°C, C_{CRIT} is 0.0082 mmol/liter, whereas for *S. marcescens* a higher value of approximately 0.015 mmol/liter is recorded for a culture temperature of 31°C.

The critical oxygen tension reported for *K. pneumoniae* grown on glucose in batch culture is in the range 8–15 mm Hg (Phillips and Johnson, 1961). In continuous culture, a slightly lower range of 2–10 mm Hg was observed (Harrison and Pirt, 1967). Reducing the DOT to below 2 mm Hg resulted in a sudden increase in the specific respiration rate of the chemostat culture. Irregular fluctuations in the DOT between 0 and 13 mm Hg were observed despite a constant rate of supply of oxygen. The higher levels of dissolved carbon dioxide resulting from enhanced CO_2 production in this "transient state" were not found to be responsible for the fluctuations in the respiration rate. Optimal diol production occurred when the DOT of the chemostat culture was below 1 mm Hg. Sparging of the culture with 1.7% oxygen was used to achieve such low DOT levels. The gas flow rate and culture working volume were not reported.

Sablayrolles and Goma (1984) used an 18-liter bioreactor sparged with 0.33 VVM air to examine the effect of K_La on the dissimilation of glucose by *K. pneumoniae*. When the oxygen absorption coefficient was set at 100 hr^{-1}, diol yields were found to be a function of the initial substrate concentration. Yields of 0.32, 0.32, 0.42, and 0.45 g/g were obtained from 22, 45, 107, and 195 g/liter initial glucose, respectively; the corresponding diol productivities were 0.76, 1.16, 2.02, and 1.10 g/liter/hour, respectively. Under the conditions employed, diol production was thus most efficient at the highest glucose concentration (resulting in an accumulation of 88 g/liter diol), while maximum productivity occurred at intermediate substrate levels. With a glucose concentration of 45 g/liter, the optimal $K_L a$ for diol production was found to be between 50 and 100 hr^{-1}. Diol yields when the initial $K_L a$ was set at 30, 100, 150, and 300 hr^{-1} were 0.44, 0.32, 0.20, and 0.17 g/g, respectively. Yields of biomass were markedly increased at the higher $K_L a$ conditions (0.27 g/g at 300 hr^{-1} versus 0.067 g/g at 30 hr^{-1}). The opposite trend was observed for the maximum specific rate of diol production (ν_{max}). The value of ν_{max} at a $K_L a$ of 30 hr^{-1} (0.70 hr^{-1}) is double that at a $K_L a$ of 300 hr^{-1}. An empirical relationship was developed by Sablayrolles and Goma for the prediction of diol yields based only on the initial glucose concentration and the oxygen absorption coefficient.

Vollbrecht (1982) has used "relative respiration rate" to measure the effect of aeration on the metabolism of gluconate by K. pneumoniae. This term is defined as the ratio of the rate of oxygen uptake by the culture during product formation to the maximum possible uptake rate (at the end of the exponential growth phase). At a ratio of 1.0, the oxygen demand of the cells is fully satisfied. The ratio is thus an index of aeration effectiveness and is especially useful at very low dissolved oxygen concentrations.

In a 4-liter bioreactor, K. pneumoniae was grown under aerobic conditions (0.10 VVM) until a cell density of 2 g/liter was achieved (Vollbrecht, 1982). At this point, gluconate was added and the aeration rate was reduced. Diol production was seen to occur at relative respiration rates between 0.05 and 0.20. Optimal production occurred at 0.10. Acetoin was also maximally produced at this ratio. Higher aeration rates resulted in the formation of organic acids. The optimal relative respiration rates for acetate, formate, and succinate were found to be 0.17, 0.17, and 0.20, respectively. The aeration conditions required to achieve these respiration rates were not reported.

Tsao (1978) examined the influence of DOT (recorded as the percentage of the medium's air saturation level) on product yields from 5% xylose. As the DOT was increased from 10% to between 20 and 30%, diol yields were reduced from 0.345 to 0.147 mol/mol of xylose consumed. Acetoin levels, however, increased correspondingly from 0.062 to 0.256 mol/mol, so that the net result was no real change in the combined yield of the two solvents (0.407 versus 0.403 mol/mol). The lowest aeration rate investigated, 10% DOT or 16 mm Hg, was greatly in excess of that reported by Harrison and Pirt (1967) to be optimal for diol production from glucose. This may in part account for the poor combined yields of diol and acetoin obtained (less than 50% of the theoretical value). An unusual observation, in marked contrast to typical results, was a doubling of the yield of ethanol from xylose at the higher aeration level (0.063 versus 0.031 mol/mol).

The effect of aeration (expressed as the oxygen supply rate; $K_L aC^*$) on the metabolism of xylose was investigated by Jansen et al. (1984). Using an initial xylose level of 20 g/liter at a controlled pH of 5.2 and temperature of 37°C, optimum diol productivity (0.44 g/liter/hour) was observed at an oxygen supply rate of 0.014 mol/liter/hour. The diol yield at this aeration rate was, however, restricted to 44% of the theoretical value. Slight improvements in yield were obtained at lower aeration rates (53% yield at 0.007 mol/liter/hour) at a cost of 10% of the maximum observed productivity.

A summary of the effects of aeration on the production of butanediol from glucose and xylose is presented in Table X. With glucose as substrate, reasonable yields appear to be achievable regardless of the aeration conditions. Diol production in anaerobic environments results in 70–74% of theoretical yields, whereas under aerobic conditions somewhat higher yields (82–90% of the theoretical value are recorded. Recall, however, that the influence of aeration is dependent on the substrate concentration, and that the results in Table X for both anaerobic and finite air conditions deal with relatively low glucose concentrations. The main effect of aeration on the dissimilation of glucose appears to be on the rate of diol formation. In the absence of oxygen, productivities of 0.08–0.14 g/liter/hour were observed; in contrast, rates as high as 2.02 g/liter/hour were recorded under aerobic conditions. The wide range of productivities noted for aerobic cultures is most likely attributable to two main factors: the initial glucose concentration and the reaction temperature employed in the individual studies.

With xylose as the carbon source, both the yield and the productivity of diol appear to be strictly dependent on the presence of environmental oxygen. Less than 10% of the theoretical yield of diol was obtained from xylose under anaerobic conditions. Poor biomass formation in the absence of oxygen appears to be the major barrier to efficient xylose conversion. This is indicated by the fact that, under finite air conditions (in which the available oxygen would be rapidly exhausted by cell synthesis), the diol yield was equivalent to those obtained in fully aerobic environments. Maximum observed yields (in the absence of medium supplements) were, however, less than 60% of the theoretical value. Diol productivity from xylose approached the lower range of the rates of formation from glucose under aerobic conditions.

H. Water Activity

Water activity (a_w) is related to osmotic pressure and varies inversely with solute concentration. It is dependent upon the molar concentration

TABLE X

EFFECT OF AERATION ON BATCH CONVERSION OF GLUCOSE AND XYLOSE

Substrate		Reactor[a]	Aeration	Temp. (°C)	pH[b]	Solvent yields (g/liter)				Diol (+ AMC)		Reference
Sugar	(g/liter)					Diol	AMC	EtOH	HAc	(g/g CHO available)	Productivity (g/liter/hr)	
Glucose	107	18-liter F	0.33 VVM	35	—	44.5	NR[c]	2.0	NR	0.42	2.02	Sablayrolles and Goma (1984)
Glucose	195	18-liter F	0.33 VVM	35	—	88.0	NR	2.4	NR	0.45	1.10	Sablayrolles and Goma (1984)
Glucose	100	19-liter jar	[d]	30	6(C)	42.0	0.0	NR	NR	0.42	0.52	Olson and Johnson (1948)
Glucose	100	19-liter jar	[e]	30	6(C)	40.5	0.0	NR	NR	0.41	0.92	Olson and Johnson (1948)
Glucose	40	WS vial	Finite air	30	6.5(I)	8.2	NR	1.9	0.2	0.21	0.17	Yu and Saddler (1982b)
Glucose	20	4-liter Erl.	Anaerobic	30	<6.3	7.4	0.1	2.7	0.2	0.37	0.08	Mickelson and Werkman (1938)
Glucose	10	WS vial	Anaerobic	37	—	3.5	NR	1.8	0.1	0.35	0.14	Yu and Saddler (1982a)
Xylose	50	7-liter F	1.10 VVM	37	5.2(A)	12.6	1.1	1.0	0.9	0.28	—	Jansen et al. (1984)
Xylose	20	7-liter F	[f]	37	5.2(A)	4.4	NR	NR	NR	0.22	0.44	Jansen et al. (1984)
Xylose	20	7-liter F	[g]	37	5.2(A)	5.3	NR	NR	NR	0.26	0.40	Jansen et al. (1984)
Xylose	10	WS vial	Finite air	37	—	2.7	NR	1.0	1.0	0.27	0.11	Yu and Saddler (1982a)
Xylose	10	WS vial	Anaerobic	37	—	0.4	NR	1.1	0.9	0.04	0.02	Yu and Saddler (1982a)

[a] F, Fermentor; jar, glass jar; WS vial, Wheaton serum vial; Erl. Erlenmeyer flask.
[b] I, Initial pH; A, automatic pH control; C, medium contains $CaCO_3$.
[c] NR, Not reported.
[d] 0.59 VVM air during the first 12 hours; 0.39 VVM during the next 24 hours; 0.20 VVM during the final 24 hours.
[e] Same as d, except using aerated inoculum.
[f] $K_{L}aC^* = 0.014$ mol/liter/hour.
[g] $K_{L}aC^* = 0.007$ mol/liter/hour.

and the activity coefficient of each solute present in an aqueous solution (Pirt, 1974). In organisms such as K. pneumoniae which are known to possess relatively weak osmotolerance (Scott, 1953), water activity may be an extremely important environmental consideration. The problems associated with the bioconversion of complex substrates such as starch or wood hydrolysates may therefore result, in part, from the high solute concentrations of such feedstocks.

Esener et al. (1981b) have demonstrated that a number of key kinetic and bioenergetic parameters are influenced by the culture water activity. Included are the duration of culture lag phase, maximum specific growth rate, biomass to substrate yield coefficient ($Y_{x/s}$), thermodynamic efficiency, and the maintenance energy coefficient. At a water activity of 0.985, growth of K. pneumoniae is reduced to one-half its optimum level. A further decline to a value of 0.975 results in a limitation of biomass formation to less than 10% of its potential (Esener et al., 1981b).

I. Inoculum

The size of the inoculum used in the bioconversion of either glucose or xylose does not appear to have any significant effect on the final yield of butanediol. Increasing the B. polymyxa inoculum from 2.5 to 5% (w/v) did not affect either the consumption rate or the diol yield from 1% xylose (Laube et al., 1984a). With K. pneumoniae, the initial fermentation rate of high xylose concentrations was influenced by the inoculum size, but the eventual diol yield was not altered (Yu and Saddler, 1982b).

Acclimatization of cultures has been shown in several cases to be extremely beneficial for the improvement of both fermentation rates and diol yields. Olson and Johnson (1948) observed that aeration of the inoculum used in the batch conversion of 10% glucose resulted in significant increases in the productivity of the process (0.92 g/liter/hour versus 0.52 g/liter/hour with an unaerated inoculum). The final yields of the diol were not, however, markedly affected. Adaptation of K. pneumoniae to high substrate concentrations prior to inoculation proved effective in the aerobic dissimilation of 150 g/liter xylose (Yu and Saddler, 1983). Under finite air conditions, enhanced biomass formation was observed with the acclimatized cultures but diol yields were not improved.

Perhaps the greatest benefits of inoculum acclimatization are to be obtained in the bioconversion of natural substrates. Concentrated hydrolysates of wood and agricultural residues are traditionally difficult to utilize. While the inhibitory agents found in these materials have not as yet been fully identified (Mes-Hartree and Saddler, 1983), breakdown products of pentose sugars (furfural, hydroxymethylfurfural, etc.) and lignin are believed to be the principal components. The high solute concentrations that are characteristic of these substrates may be a major factor in the inhibition of such poorly osmotolerant species as K. pneumoniae. Perlman (1944) found that supplementation of southern red oak hydrolysate with malt sprouts was not required if the K. pneumoniae inoculum was first acclimatized to the substrate. Diol yields from the hydrolysate (100 g/liter initial carbohydrate) with an unacclimatized culture, with an unacclimatized culture plus malt sprout supplement, and with an acclimatized culture were 0.17, 0.35, and 0.36 g/g, respectively. Diol productivities were 0.34, 0.73, and 1.04 g/liter/hour, respectively.

V. Potential Substrates

Industrial-scale production of fuels or chemical feedstocks via bioconversion technology requires the availability of substrates that are both abundant and inexpensive. Several industrial waste resources are felt to offer considerable potential. Included are agricultural and logging residues, pulp and paper mill effluents, food industry wastes, and molasses. The production of cellulosic biomass specifically for bioconversion purposes has also been suggested. Thus the possibilities of silviculture, and sugar crop (e.g., sugar beet) production.

Those natural substrates that have been shown to offer the greatest potential for the industrial manufacture of butanediol are discussed briefly below. Diol yields and productivities obtained from the conversion of these feedstocks are presented in Table XI.

A. Waste Sulfite Liquor

The waste liquor produced from sulfite pulp and paper mills poses a serious disposal problem for the industry. Annual worldwide production of this effluent is estimated at approximately 100×10^6 metric tons (Forage and Righelato, 1979). Its composition is variable, depending upon the species of wood pulped and the conditions employed. Analysis of the liquor from spruce pulping shows that hemicellulosic carbohydrates account for approximately 35% of the organic dry material present (Detroy and Hesseltine, 1978).

Pretreatment of the liquor prior to microbial cultivation usually includes stripping of sulfur dioxide by combined boiling and aeration. In

TABLE XI

SUMMARY OF DIOL PRODUCTION FROM POTENTIAL SUBSTRATES

Substrate[a]	Initial monosaccharide		Ferm. time (hr)	Yields (g/liter)			Butanediol			Ref.[b]
	g/liter	%Used		Diol	EtOH	HAc	g/g used	g/g available	g/liter/hr	
Waste sulfite liquor	38.0	69.7	72	9.0	3.3	—	0.34	0.24	0.13	1
Citrus press juice	215.0	91.5	56[b]	51.0	—	—	0.26	0.24	0.91	2
Sugar beet molasses	56.6	71.0	24	20.1	—	—	0.50	0.36	0.84	3
Sugar beet pulp	11.0	78.0	11	2.4	1.8	—	0.28	0.21	0.21	4
Wood hydrolysate (1a)	100.0	55.0	48	16.5	—	—	0.30	0.17	0.34	5
Wood hydrolysate (1b)	100.0	93.0	34	35.5	—	—	0.38	0.36	1.04	5
Wood hydrolysate (2)	12.1	95.0	NR[c]	6.0	3.8	2.1	0.52[d]	0.50[d]	—	6
Wood hydrolysate (3)	9.7	100.0	24	0.2	0.5	3.8	0.02[d]	0.02[d]	0.01	7
Wood hydrolysate (4)	40.0	100.0	48	20.0	5.9	0.1	0.50[d]	0.50[d]	0.42	8

[a] Wood hydrolysate: (1a), southern red oak hydrolysate (Scholler process); (1b), same as 1a, except for acclimatized culture; (2), steam-exploded aspen; hemicellulose fraction; acid hydrolysis; (3), steam-exploded aspen; hemicellulose fraction; enzyme hydrolysis; (4), steam-exploded aspen; cellulose fraction; acid hydrolysis.
[b] Average value.
[c] Not reported.
[d] Conversion of HAc and uronic acids not considered.
[e] References: (1), Murphy and Stranks (1951); (2), Long and Patrick (1961); (3), McCall and Georgi (1954); (4), Blanco et al. (1984); (5), Perlman (1944); (6), Yu et al. (1982); (7), Saddler et al. (1983); and (8), Yu et al. (1984a).

studies conducted by Murphy and Stranks (1951), dibasic ammonium phosphate (0.1%) and 50 ml of 10% molasses per liter of medium were added as supplementary nutrients before neutralization of the liquor with ammonium hydroxide. Calcium carbonate (2%) was added prior to sterilization at 15 psi for 10 minutes. Results of the bioconversion of this substrate are given in Table XI. Approximately 70% of the initial sugar content (38 g/liter) was consumed in 72 hours, resulting in the accumulation of 9.0 g/liter diol (0.24 g/g available) and 3.3 g/liter ethanol. Productivity, at 0.13 g/liter/hour, was relatively poor. *Aeromonas hydrophila* was found to be slightly more efficient in the utilization of this substrate, giving a butanediol yield of 12.4 g/liter (0.33 g/g) under similar conditions. Productivity was 0.17 g/liter/hour. Ethanol formation (3.9 g/liter) was also slightly enhanced with *A. hydrophila*.

B. Food Industry Wastes

Large-scale utilization of food industry waste for microbial conversion is, in most cases, considered to be improbable (Kosaric *et al.*, 1983). Seasonal availability, high variability of composition, disperse origin, and extensive competition from established by-product and feed markets are seen to be major obstacles.

Although from an economic standpoint, the potential for industrial fuel and chemical production from food waste appears to be limited, its technical feasibility has been demonstrated in several cases. Long and Patrick (1961) observed reasonable diol productivities (0.91 g/liter/hour) when *K. pneumoniae* was grown on concentrated citrus canning-plant press juice (215 g/liter initial carbohydrate). The conversion efficiency was, however, less than 50% of the theoretical value (Table XI). Supplementation of the substrate (obtained from the grinding, liming, and pressing of citrus peel) with potassium phosphate and urea was found to be necessary. Acclimatization of *B. polymyxa* to citrus molasses did not result in improved diol yields (Long and Patrick, 1965). Fermentation times were, however, reduced. In a 15-liter fermentor aerated at a rate of 0.11 VVM, the conversion of 175 g/liter total sugar resulted in the accumulation of 42 g/liter diol at a productivity of 0.105 g/liter/hour. Cooper (1976) has shown that although considerable quantities of this particular resource are generated by the citrus industry (3.6×10^6 tons/year, 97% is consumed in the manufacture of by-products.

The annual production of cheese whey in 1974 was estimated to be, on a global basis, approximately 74×10^6 tons (FAO, 1974). Approximately 40% of the 16×10^6 kg of whey produced yearly in the United

States is dumped into waterways and sewage systems (Kosikowski, 1979). The high biological oxygen demand (BOD = 60,000–70,000 mg/liter) of this material poses serious environmental problems.

Speckman and Collins (1982) tested several microorganisms for diol production from lactose (the principal carbohydrate in cheese whey) and observed positive results with *B. polymyxa*. When grown on sweet whey containing 34.2 g/liter lactose, the diol level after 7 days at 30°C was 22.2 g/liter. This represents a yield of 0.25 g/g of lactose consumed, or 0.16 g/g available. Diol productivity was 0.03 g/liter/hour.

Lee and Maddox (1984) found that *K. pneumoniae* was also capable of butanediol production from lactose. When cultivated on whey permeate (the material left after ultrafiltration-based recovery of soluble protein from whey), yields of 0.46 g/g of lactose consumed were reported. The final diol concentration achieved was 7.5 g/liter. The initial lactose content of the medium was not given. Ethanol and acetoin levels were 0.3 and 1.1 g/liter, respectively. The combined productivity of diol and acetoin was 0.09 g/liter/hour. Hydrolysis of the permeate by a fungal lactase preparation prior to inoculation with *K. pneumoniae* led to accululation of 13.7 g/liter diol from an initial carbohydrate content (lactose, glucose, and galactose) of 38.8 g/liter. Diol yields were thus 0.35 g/g of sugar available. Productivity was 0.14 g/liter/hour.

C. Molasses

The utilization of molasses for the production of 2,3-butanediol has been examined by numerous investigators. Blackstrap molasses (the residue left after the crystallization of sucrose from sugar cane juice) and citrus molasses have been tested by Owen (1950), Long and Patrick (1960), and Bahadur and Rangenayaki (1960). Sugar beet molasses has been studied by McCall (1943), McCall and Georgi (1954), Freeman and Morrison (1947), and Anastassiadis and Wheat (1953). Very little nutrient supplementation is required for the conversion of molasses. The material can be stored in its concentrated form for long periods and diluted to the optimum sugar content prior to use. Strong competition exists, however, for this substrate, such that its utilization for diol production is generally considered to be uneconomical. Pilot- and commercial-scale processes for diol production from beet molasses are examined in Section VIII,A.

McCall and Georgi (1954) examined diol production from sugar beet molasses diluted to between 4 and 22% sucrose and supplemented with 1% ground barley malt. Under the conditions employed, 67% of the theoretical yield of diol was obtained in 24 hours when the initial sucrose content was 57 g/liter. The overall productivity was thus 0.84 g/liter/hour. The yield after 72 hours from 164 g/liter sucrose was 42% of the theoretical value, resulting in a productivity of 0.51 g/liter/hour.

D. Sugar Beet Pulp

The pulp remaining after the extraction of juice from sugar beets represents approximately 24% of the dry weight of the beet root (Nathan, 1978). This material has a high feed value in either wet or dry form. Despite this highly competitive application, Blanco et al. (1984) examined the conversion of beet pulp by K. pneumoniae, following enzymatic hydrolysis by Trichoderma reesei cellulase. The initial hydrolysis is necessitated by the high cellulose and hemicellulose content of the substrate (approximately 44% of the dry weight). Following supplementation of the hydrolysate with potassium phosphate, ammonium sulfate (5 g/liter), yeast extract (3 g/liter), and trace metals, the diol yield from a relatively dilute carbohydrate content of 1.1% was limited to 2.4 g/liter (approximately 43% of the theoretical value). Ethanol, at 1.78 g/liter was a major by-product of the fermentation. The diol productivity observed, 0.21 g/liter/hour, is quite low, and does not include the time required for the enzymatic hydrolysis of the substrate.

E. Wood Hydrolysates

While logging residues represent a vast supply of waste biomass, the problems associated with their collection from widely distributed sites are generally believed to render this source uneconomical for bioconversion. Silviculture, the intensive farming of trees, is, however, regarded to be a viable alternative to traditional sources of fermentation substates. Hardwood species are preferred for their high cellulose content and rapid juvenile growth. At present, the yield of cellulosic-based biomass is approximately 5–12 tons/acre. Projections for annual harvests of up to 20 tons/acre (dry material) should be realized through strain development and improved agricultural technology (Salo and Henry, 1979).

Perlman (1944) examined the ability of K. pneumoniae to utilize the carbohydrates present in hydrolysates from several hardwood and softwood species. Hydrolysis was performed by percolation of dilute sulfuric acid (0.2–1.0%) through wood chips at approximately 160°C (Scholler process). Following neutralization of the hydrolysate with alkali or lime, urea (4 g/liter), potassium phosphate (2 g/liter), and calcium carbonate (5 g/liter) were added. Further supplementation of the medium with yeast extract or malt sprouts (10 g/liter) was not required if cultures were employed which had been acclimatized to the hydrolysate. Accumulation of 35.5 g/liter diol was reported within 48 hours of growth (productivity: 1.04 g/liter/hour) from 100 g/liter initial reducing sugar. The yield of butanediol was thus approximately 70% of the theoretical value.

In a series of investigations, Yu and co-workers studied solvent production by K. pneumoniae grown on sugars derived from hardwood aspen. Substrate preparation was performed in two major operations. Initial steam explosion of wood chips (under various combinations of steam temperature and duration) preceded hydrolysis by either acidic or enzymatic means. Two fractions were distinguished following steam explosion: a water-soluble, hemicellulose-rich component, and a water-insoluble, cellulose-rich fraction. Each component was then subjected to acid or enzymatic hydrolysis. Flowsheets of the processing steps employed in substrate preparation are presented in Fig. 8.

When the conditions of steam explosion were selected so as to obtain the optimum yield of undegraded hemicellulose in the water extract (250°C, 15 seconds), subsequent hydrolysis with 3% sulfuric acid for 1 hour at 125°C resulted in the breakdown of 65–75% of the total reducing sugars to monosaccharides (Yu et al., 1982). D-Xylose was the major component of the hydrolysate. A flowsheet for the preparation of this hemicellulose-rich fraction is given in Fig. 8a. The hydrolysate was concentrated and added to a chemically defined medium such that the final monosaccharide content was 12.1 g/liter. Under finite air conditions, the concentrations of diol and ethanol achieved were 6.02 and 3.81 g/liter, respectively. This high diol yield (0.50 g/g monosacharide available) does not take into account the conversion af acetic and uronic acids present in the hydrolysate (0.5% acetate was also included in the medium formulation). On a molar basis, ethanol production exceeded that of butanediol (83 mM versus 67 mM.). The molar ratio of ethanol to diol was thus 1.24. The relative proportion of ethanol formation from acid-hydrolyzed xylan was markedly lower, with a molar ratio of ethanol to diol of 0.71. At hydrolysate concentrations exceeding 20 g/liter, the efficiency of bioconversion was decreased, apparently due to the presence of inhibitors in the substrate.

Enzymatic hydrolysis of the hemicellulose fraction of steam-exploded aspen (Fig. 8b) achieved a monosaccharide yield of greater than 50% of the reducing sugars (Saddler et al., 1983). While 100% utilization of the monosaccharides (1% initially) was achieved 24 hours after inoculation with K. pneumoniae, diol yields were extremely poor. The concentrations of butanediol, ethanol, and acetic acid were 0.18, 0.52, and 3.83

a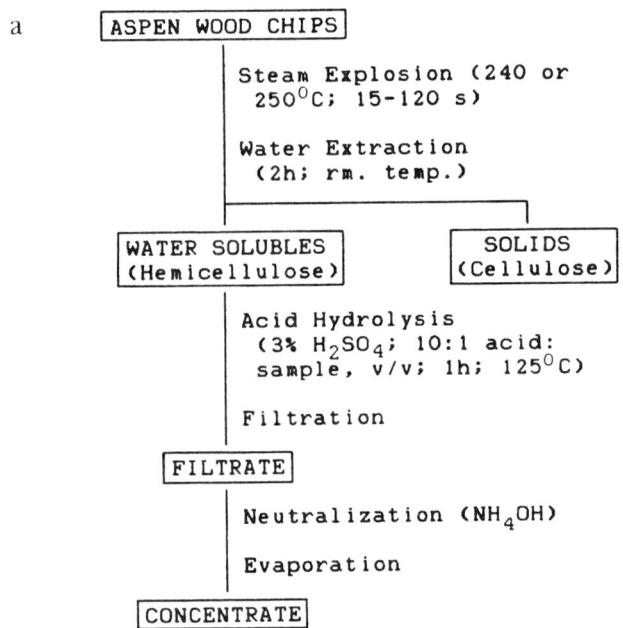

b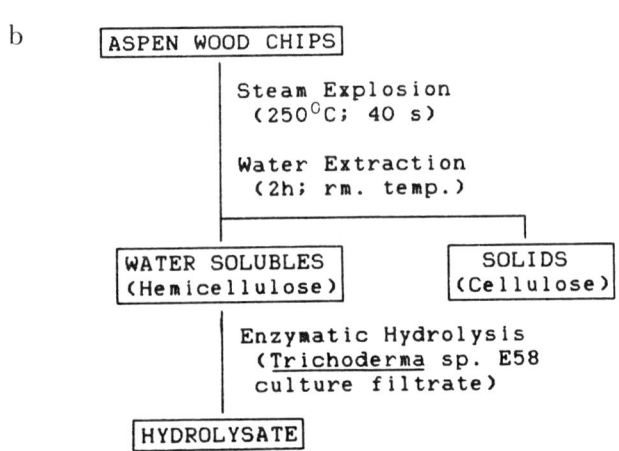

FIG. 8. (a) Preparation of hemicellulose component of aspen by combined stream explosion and acid hydrolysis. (b) Preparation of hemicellulose component of aspen by combined steam explosion and enzymatic hydrolysis. (c) Preparation of cellulose component of aspen by combined steam explosion and acid hydrolysis.

C

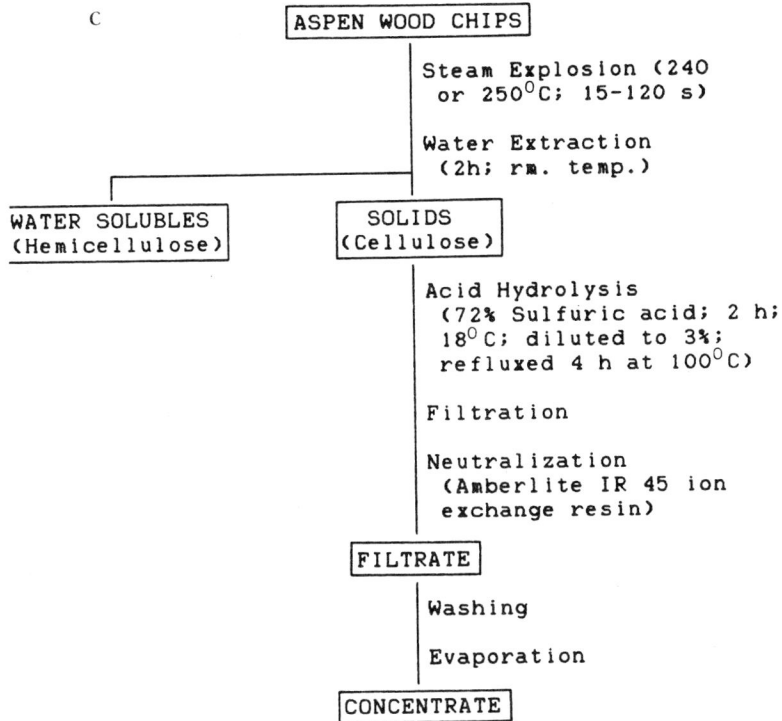

FIG. 8. *(Continued)*

g/liter, respectively. Diol formation was thus limited to approximately 4% of the theoretical yield. On a molar basis, acetate was produced at a rate 32 times greater than was butanediol.

The metabolism of the monosaccharides released from the hemicellulose fraction of steam-exploded aspen appears, therefore, to be dependent upon the method of hydrolysis employed. With acid hydrolysis, the relative production of butanediol, ethanol, and acetic acid, expressed as the mole percent of the total solvent production, is 36.2, 44.6, and 19.2%, respectively. The corresponding figures after enzymatic hydrolysis are 2.6, 14.7, and 82.7%, respectively. A marked shift to acetate production at the expense especially of butanediol is evident when the enzymatically hydrolyzed substrate was employed.

This shift may in part result from the conditions employed in the initial steam explosion of the enzymatically hydrolyzed substrate. The optimum conditions for recovery of undegraded hemicellulose from the aspen chips (those employed in the pretreatment of the acid-hydrolyzed material) were found to be 250°C for 15 seconds. Although the same temperature was employed in the steam explosion of the wood chips that were then enzymatically hydrolyzed, a considerably longer duration (40 seconds) may have led to formation of metabolic inhibitors.

Solvent extraction of the hemicellulose fraction (a 40-second steaming) with an ethanol–benzene mixture and subsequent fermentation led to a total solvent yield nearly double that obtained after enzymatic hydrolysis. The observed concentrations of diol, ethanol, and acetate were 2.2, 1.4, and 5.6 g/liter, respectively.

The optimal steaming duration at 240°C for solvent production from the acid-hydrolyzed cellulose fraction of aspen (Fig. 8c) was found to be 30 seconds (Yu et al., 1984a). Under these conditions, 75% of the original wood was recovered as water insolubles, of which 62.2% was hexosan and 8.0% was pentosan in nature. Following acid hydrolysis, the hexosan content of the substrate was 61.4% and the pentose sugars made up 3.5% of the available carbohydrate. Cultivation of this substrate (40 g/liter initial sugar) under finite air conditions resulted in the accumulation of 20.01 g/liter diol, 5.86 g/liter ethanol, and only 0.12 g/liter acetic acid. Diol productivity was 0.42 g/liter/hour.

F. AGRICULTURAL SUBSTRATES

Several agricultural residues have been considered as possible cellulosic feedstocks for the production of liquid fuels. Yu et al. (1984b) examined the production of diol by *K. pneumoniae* grown on hydrolysates of steam-exploded barley straw, corn stalks, and wheat straw. Yields of diol, from initial substrate concentrations of 20 g/liter, were 2.97, 2.60, and 2.66 g/liter, respectively. Ethanol formation accounted for an additional 0.83, 0.67, and 1.59 g/liter, respectively. In these trials, solvent production and enzymatic hydrolysis by xylanases from *Trichoderma harzianum* E58 were simultaneously conducted in a single vessel.

Research on diol production from cereal grains was carried to the pilot-plant scale in the late 1940s. The amylolytic bacterium *B. polymyxa* was employed to convert starchy substrates such as wheat (Wheat et al., 1948) and barley mashes (Tomkins et al., 1948). Aspects of the pilot and proposed commercial operations are discussed in Section VIII.

VI. Culture Techniques

Environmental factors such as temperature, pH, aeration, and medium composition have been shown to significantly influence the production of butanediol by bacterial species. The operational conditions of the fermentation vessel are similarly important in the establishment of an optimal process design. Thus the effects on diol yield and productivity of batch versus continuous modes, and free cell versus immobilized systems, must be evaluated. A summary of the effects of reactor operation mode on butanediol formation is presented in Table XII.

While a concentrated product stream is desirable in any bioconversion scheme, it is essential in the production of butanediol due to the difficulties encountered in diol recovery (see Section VII). A minimum diol concentration of approximately 80 g/liter has been estimated to be required for economically feasible recovery (Jansen, 1982). This has led several researchers to believe that successful production of this compound on a commercial scale will require a batch mode of operation.

In simple batch culture at an aeration rate of 0.33 VVM, Sablayrolles and Goma (1984) achieved diol yields of 88 g/liter from 195 g/liter initial glucose. Productivity could be nearly doubled (from 1.10 to 2.02 g/liter/hour) by reducing the initial level of glucose to 107 g/liter. Diol yields of 99 g/liter were obtained by Olson and Johnson (1948) through the slow addition of a concentrated solution of glucose (plus potassium phosphate and urea) to a batch culture of K. pneumoniae. Initially 5.9 g/liter, the concentration of glucose in the culture was maintained at 3 g/liter for the duration of the experiment. The culture was aerated at a rate of 0.59 VVM during the initial 14 hours of growth, and at 0.39 VVM thereafter. Production of acetoin accounted for an additional 9.3 g/liter. The combined yield of acetoin and butanediol was thus 82% of the theoretical value after 108 hours of growth, giving a productivity of 1.00 g/liter/hour (0.92 g/liter/hour for the diol alone).

Similar yields of diol were obtained by Yu and Saddler (1983) using a double-fed batch approach. Daily additions of carbohydrate and yeast extract were made to raise the sugar content of the medium by 20 g/liter. The employment of cultures of K. pneumoniae that had been acclimatized to high substrate concentrations resulted in the accumulation of 106 g/liter diol from 225 g/liter glucose, and 81 g/liter diol from 189 g/liter xylose. When high initial substrate concentrations (150 g/liter were used, the combined yields of acetoin and diol were 113 and 88 g/liter, respectively, from 226 g/liter glucose and 190 g/liter xylose.

Continuous cultivation techniques are often favored in biomass conversion due to their inherent capacity for elevated productivities. Rates of diol production of 4.60 g/liter/hour have been reported by Pirt and Callow (1958) from a 100-g/liter sucrose feed. The diol concentration in the effluent, was, however, only 23 g/liter. Reducing the dilution rate

TABLE XII

EFFECT OF REACTOR OPERATION MODE ON DIOL PRODUCTION

Substrate							Butanediol			
Sugar	(g/liter)	Reactor[a]	Reactor mode[b]	Temp. (°C)	pH		g/liter	g/g	g/liter/hr	Ref.[c]
Glucose	195	18-liter F	Batch	35	—		88.0	0.45	1.10	1
Glucose	107	18-liter F	Batch	35	—		44.5	0.42	2.02	1
Glucose	265	19-liter jar	SF	30	6.0		99.0	0.37	0.92	2
Glucose	225	S flask	DFB	37	6.5		106.0	0.47[d]	0.74	3
Xylose	189	S flask	DFB	37	6.5		81.0	0.43[d]	0.56	3
Sucrose	100	2-liter F	Cont. (1)	35	5.5		29.8	0.30	3.28	4
Sucrose	100	2-liter F	Cont. (2)	35	5.5		23.0	0.23	4.60	4
Sucrose	140	a	Cont. (3)	37/30	5.3		67.5	0.48	2.58	5
Glucose	50	S flask	IB	37	7.0		15.0	0.30	0.50	6
Glucose	25	Flask	IC	37	7.0		3.0	0.12	0.75	6

[a] F, Fermentor; S, shake; a = 0.5-liter fermentor first stage, 4-liter fermentor second stage.
[b] SF, Slow feed; DFB, double-fed batch. Continuous modes: Cont. (1), D = 0.11 hr^{-1}; Cont. (2), D = 0.20 hr^{-1}; Cont. (3), two-stage continuous (first stage D = 0.214 hr^{-1}, second stage D = 0.50 hr^{-1}; IB, immobilized batch; IC, immobilized continuous.
[c] References: (1), Sablayrolles and Goma (1984); (2), Olson and Johnson (1948); (3), Yu and Saddler (1983); (4), Pirt and Callow (1958); (5), Pirt and Callow (1959); (6), Chua et al. (1980).
[d] Conversion of acetic acid in medium was not considered.

from 0.20 to 0.11 hr^{-1} resulted in an increase in the diol level to 30 g/liter at the expense of productivity (3.28 g/liter/hour). Further increases in the diol concentration were made possible by adoption of a two-stage continuous culture approach (Pirt and Callow, 1959). The initial reactor was operated under conditions promoting the formation of biomass (37°C; dilution rate of 0.214 hr^{-1}; aeration rate so as to achieve a culture oxygen-uptake rate of 46.3 mmol/liter/hour). The output of this stage was used to feed a second, larger reactor (30°C; dilution rate of 0.050 hr^{-1}; oxygen uptake rate of 10.2 mmol/liter/hour). The concentration of the diol obtained in the final product was 67.5 g/liter (plus 3.0 g/liter acetoin). The overall diol productivity of the two-stage system was 2.58 g/liter/hour. Continuous formation of butanediol at concentrations suitable for economic recovery does, therefore, appear to be a reasonable objective.

Production of butanediol by immobilized cultures has had limited success to date. Batch cultivation of K. pneumoniae immobilized in $\varkappa$-carrageenan resulted in the accumulation of 15 g/liter diol from 50 g/liter glucose at a productivity of 0.50 g/liter/hour (Chua et al., 1980). A 50% improvement in productivity resulted from a switch to a continuous mode of operation, but the resultant yield of diol from the 25-g/liter glucose feed, at 3.0 g/liter, was only 24% of the theoretical value.

VII. Butanediol Recovery

The physical and chemical properties of 2,3-butanediol (high boiling point, hygroscopicity, etc.), make its recovery from fermentation broths an extremely difficult task by conventional methods. Dilution of the broth with high-boiling liquids such as glycerol has been found to facilitate distillation of the diol. Spray drying, in which volatile fermentation products would become vaporized upon spraying the broth into a stream of high-temperature inert gas, was proposed by Liebmann in 1945. Separation of the solids (for possible use as animal feed) would precede condensation of the products. Recovery based on chemical conversion of the diol has been suggested by several researchers. Senkus (1946) advocated reaction of the diol with formaldehyde, separation of the formal by distillation, then acid treatment to recover the diol. A similar procedure using butyraldehyde was developed by Tink in 1950.

Solvent extraction has been found to be an effective means of diol recovery. Capable of selectively recovering the diol from dilute solutions, solvent extraction can achieve separation of the diol from water-soluble impurities such as sugars and proteins. This enables the isolation of a purer end product, as well as reducing equipment fouling in subsequent evaporation and distillation operations. Suitable solvents include diethyl ether (Birkinshaw et al., 1931; Kolfenbach et al., 1944) and n-butyl alcohol (Othmer et al., 1945). Tsao (1978) found that 75% of the diol present in a fermentation broth could be recovered by a single extraction with diethyl ether. Recovery of coproducts acetoin, ethanol, and diacetyl was found to be 65, 25, and 75–90%, respectively. Othmer et al. (1945) proposed process schemes for the recovery of butanediol from fermented grain mashes by extraction with either n-butyl alcohol or butanediol diacetate. The initial stages of product recovery were identical for the two methods (Fig. 9). Following pH adjustment with lime to between 8.0 and 8.5, the mash was heated, filtered, and concentrated in a triple-effect evaporator with backward feed. A small distillation column on the first effect was incorporated to prevent the diol from distilling over. Ethanol was removed in the overhead from the final effect and was concentrated by steam distillation. The concentrated diol solution ($\sim 20\%$ w/v; 70°C) was fed to a single-stage continuous-countercurrent liquid–liquid extractor. When butanol was used as the extracting solvent (Fig. 10), the extract layer was found to contain essentially all of the diol and approximately 30% of the water introduced to the unit. Butanol was recovered from the extract by steam distillation. The diol-enriched bottoms from the solvent recovery column were then fed to an esterifying column, to which acetic acid (plus a small amount of sulfuric acid catalyst) was also introduced near the feed plate. Butyl acetate, added to the column as an entrainer, removed the water formed by the esterification of the diol. The butanediol diacetate formed in the process was recovered from the base of the column and was neutralized with sodium acetate.

The equipment required for butanediol diacetate-based extraction of the evaporator product (Fig. 11) is simplified by the elimination of an extraneous extracting solvent. The diol-containing extract layer is fed directly to the esterifying column, where the small amount of dissolved water ($\sim 8\%$ w/v) is removed by an entrainer. The acetylated product is neutralized with sodium acetate after a portion has been recycled to the extractor. The diacetate formed in this process can be readily converted to 1,3-butadiene (refer to Section VIII,D) for subsequent manufacture of synthetic rubber.

Weizmann et al. (1948) noted that 2,3-butanediol was easily adsorbed on active carbon, and proposed that this operation precede solvent extraction. Diol recoveries were reported to be 100% when acetone or dioxane was used as the extracting agent in a Soxhlet-like apparatus. A

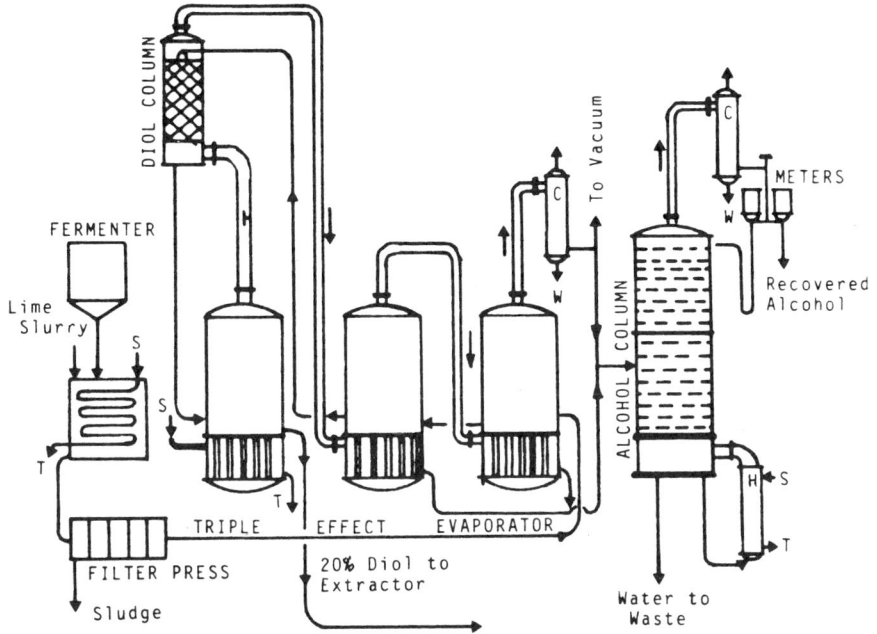

FIG. 9. Flowsheet for preliminary treatment of fermentation liquors containing 2,3-butanediol (source, Othmer et al., 1945).

current of warm air passed over the adsorbate resulted in fairly complete recoveries, but only 10% of the product was found as diol, the balance as acetylmethylcarbinol and diacetyl. It was speculated that heavy metals present in the active carbon acted as catalysts for the dehydrogenation reactions observed.

The most practical methods of diol recovery appear to involve countercurrent stream stripping. Successful pilot-scale steam-stripping operations were established in the mid-1940s by both the Northern Regional Research Laboratory, Peoria, Illinois (Blom et al., 1945) and the National Research Council of Canada (Wheat et al., 1948). Blom et al. examined diol recoveries obtained in the stripping of fermented corn and wheat mashes. Before introduction to the stripping column, the mash was concentrated to 13.8% diol and approximately 30% solids in an evaporator. The column consisted of a 0.15-m-diameter jacketed pipe, 3.96 m long and filled with 6.4-cm ceramic spheres to a height of 3.51 m. With the concentrated mash fed into the top of the column, and steam at 758 kPa (gauge) supplied to the base, a diol recovery of 95%

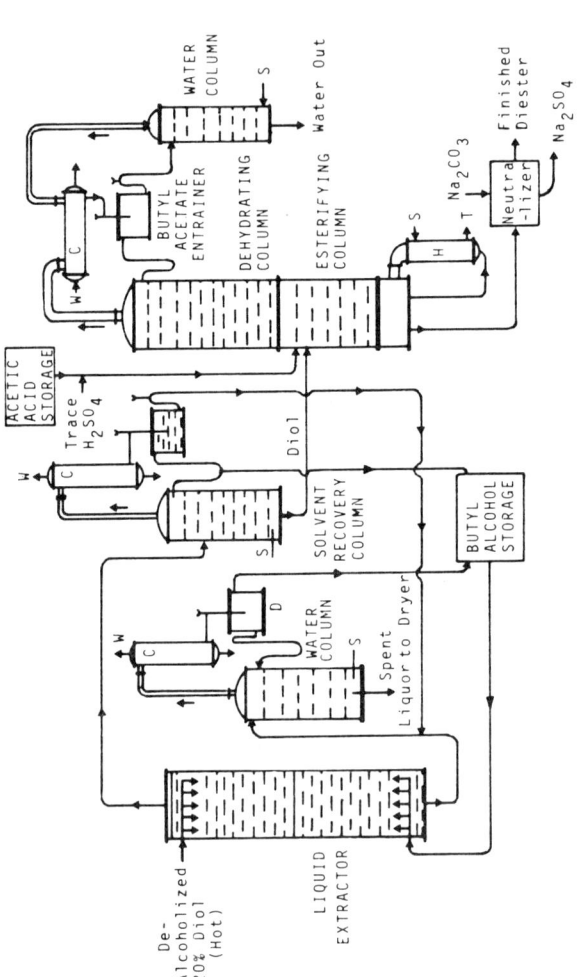

FIG. 10. Flowsheet for process using n-butanol as solvent in the recovery of 2,3-butanediol (source, Othmer et al., 1945).

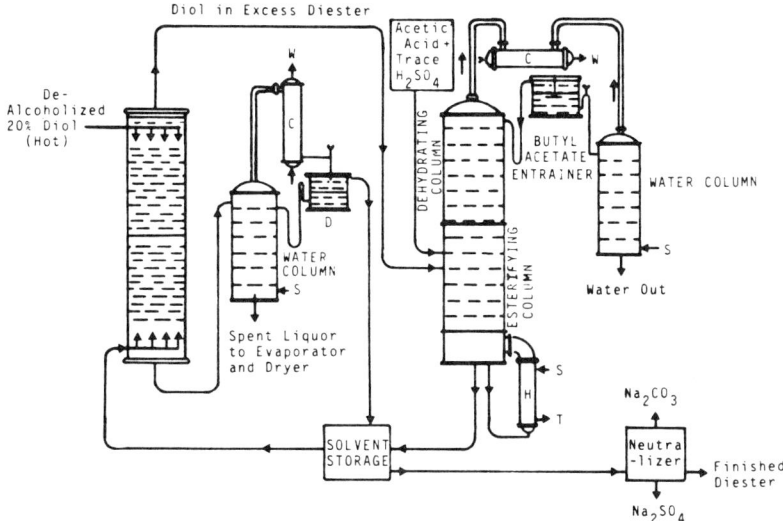

FIG. 11. Flowsheet for process using butanediol diacetate as solvent in the recovery of 2,3-butanediol (source, Othmer et al., 1945).

could be obtained with the use of 6.08 parts of steam per 1 part of syrup. When the column was extended to 8.23 m and packed randomly with iron Raschig rings (4.8 by 4.8 cm), rapid fouling of the column was observed. Careful packing, coupled with cyclic washing of the column, was therefore recommended. The sludge which formed on the packing was softer and more easily removed by water washing if the syrup was made alkaline by the addition of 1.25% calcium hydroxide.

VIII. Process Design

Processes for the pilot- and commercial-scale production of butanediol from molasses, wheat, and barley are described below. A brief review of proposed operations for the generation of 1,3-butadiene from the diol is also provided. (See Section IX for revised cost estimates for commercial production of diol.)

A. Butanediol Production from Molasses

Pilot-scale production of 2,3-butanediol was examined by Wheat in 1953 using sugar beet molasses as substrate. The 4000-liter batches were prepared by dilution of approximately 540 kg of molasses (initial sugar content, 51–57%) to achieve a final sugar concentration of approximately 7.25% (Wheat, 1953a). Following combination of the molasses, water, phosphate, and steam in a 230-liter slurry tank, the resultant mash was heated by mixture with steam at 1480 kPa in a jet heater. The mash was then passed through a pressurized continuous cooker (45.7 m of 2.54-cm Monel pipe). Residence time in the cooker was 1.7 minutes, inlet and outlet temperatures being 175 and 163°C, respectively. The cooked mash was introduced tangentially into a flash cooler (Monel cylinder, 0.61 m in diameter, 0.61-m high with a 60° conical bottom). Fed by gravity to the 6800-liter fermentor, the pH of the mash was adjusted to 6.2 by the addition of 80% acetic acid. Following inoculation with *K. pneumoniae*, the medium was aerated at a rate of approximately 0.02 VVM for the first 24 hours of growth and 0.01 VVM for the next 12 hours. Aeration was discontinued after 36 hours. Sugar utilization ranged between 95.9 and 99.9%, with a fermentation efficiency of 82.6 to 96.6%. The average combined yield of acetoin and butanediol was found to be 181.2 kg per 1000 kg of molasses. Ethanol yields were recorded in the range 25.5 to 65.1 kg per 1000 kg of molasses.

Product recovery from the fermented mashes was initiated by steam stripping to remove ethanol (Wheat, 1953c). A 0.6-m-diameter column containing 15 stripping and 7 enriching plates at 43.2-cm spacings was employed for this purpose. Slops from this column were fed to a vertical tube evaporator constructed of stainless steel. The 160 tubes in the unit provided a liquid-side heat transfer area of 12.7 m^2. Vapor from the evaporator was scrubbed of diol in a 33-plate, 0.31-m-diameter copper rectifying column. When the diol concentration reached 20% (w/v), syrup was continuously removed from the base of the evaporator and fed to the top of a 6.1-m stripping column packed with 2.5-cm Raschig rings to a depth of 5.5 m. Steam at 480 kPa was introduced at the base of the 0.31-m-diameter unit. The diol-enriched vapor produced was scrubbed in a second identical column by preheated water fed into the top of the unit. The scrubber product was further purified in a two-step distillation operation. A detailed description of the design and performance characteristics of the evaporation and distillation unit was given by Wheat (1953b).

Based on the results of these pilot plant studies, a detailed design was proposed for a commercial-scale plant capable of processing 27,200 kg of molasses per day (Wheat, 1953c). The carbohydrate content of this quantity of material is approximately equal to that of the daily wheat consumption of a similar design (see Section VIII,B). Flowsheets of the fermentation and recovery sections of the molasses plant are given in Figs. 12 and 13, respectively. The total and working capacities, respectively, for each

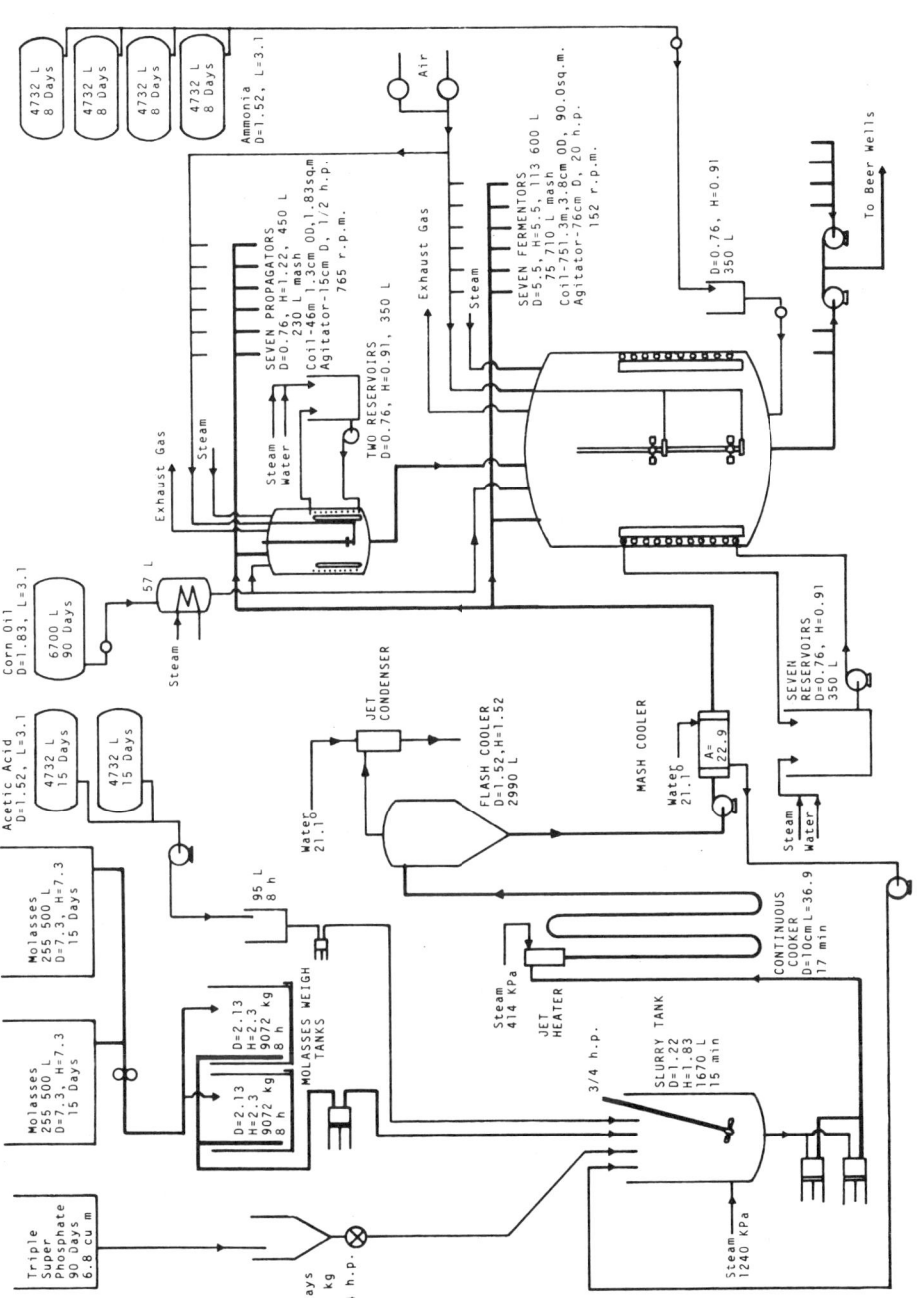

FIG. 12. Flowsheet for commercial operation processing 27 metric tons of molasses per day: Fermentation section; H = height, meters; D = diameter, meters; P = number of plates; A = heat transfer area, square meters (source, Wheat, 1953c).

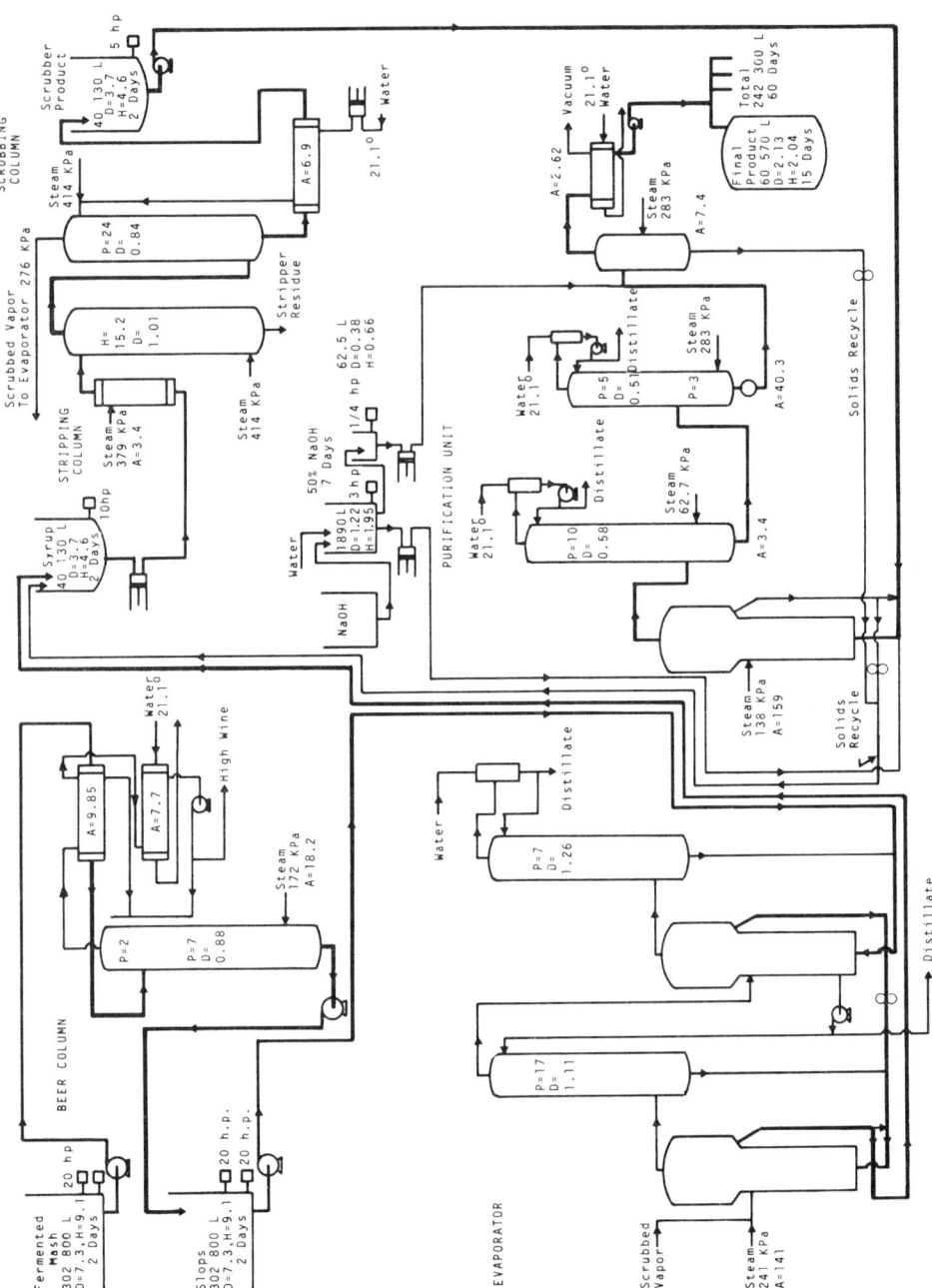

FIG. 13. Flowsheet for commercial operation processing 27 metric tons of molasses per day: Recovery section; H = height, meters; D = diameter, meters; P = number of plates; A = heat transfer area, square meters (source, Wheat, 1953c).

of the fermentors are 136,000 and 91,000 liters. At a mashing rate of approximately 182,000 liters/day, two units could be filled in a single day. With a fermentation time of 36 hours (plus 12 hours without aeration in the beer wells), a 60-hour cycle is required for each unit, and therefore a minimum of five units would be required for continuous operation. A total of seven fermentors was included in the process design. Preparation of inoculum was to occur in seven (four more than the minimum design requirements) 550-liter propagators. The utilization of 270 liters of mash for a 36-hour growth of the inoculum would provide the same ratio of inoculum to mash as was used in the pilot plant studies. A double-effect evaporator (each unit equipped with a rectifying column) was recommended for concentration of the ethanol-stripped syrup. The scale of all major plant equipment is given in Figs. 12 and 13. Flow rates for each of the process streams are provided in Table XIII. It was assumed that two molasses storage tanks would be located outdoors, and other raw materials, spare pieces of large equipment, and the final product could by stored in a building 19.5 × 10.4 × 5.5 m. A 38.4 × 16.5 × 11.0-m structure would be required to house the fermentors, beer well, and slops tank. Including structures for offices, laboratory, and locker and control rooms, a total area of approximately 37 × 67 m was considered to be sufficient for the entire operation.

Estimated costs of plant construction and daily production for the proposed commercial-scale process were calculated by Wheat (1953c) and are summarized in Tables XIV and XV. In 1952 dollars, the total cost for building the plant was calculated to be approximately $2,780,000. Daily production expenses (including management, insurance, taxes, and depreciation) were expected to total $2660. The molasses accounted for approximately 45% of the total operating costs (which do not include the fixed costs outlined in Table XV). At a daily production rate of 4712 kg of diol, the production cost was determined to be 56.4 cents/kg. The selling price required for a 20% return on the investment was $1.07/kg. Tripling the plant capacity was estimated to result in a reduction in the selling price of the diol to 81.6 cents/kg. At approximately the same time as these studies were being performed, the Celanese Corporation of America announced the availability, in tank-car quantities, of synthetically manufactured, 2,3-butanediol for a price of 30.9 cents/kg (Chem. Eng., 1951). The selling price of 1,3-butadiene was 35.3 cents/kg at that time (Chem. Eng. News, 1952).

B. Butanediol Production from Wheat

Feed wheat was used for diol production on a pilot plant scale by

TABLE XIII

MASS FLOW RATES OF PROCESS STREAMS IN A 27,200-kg/day MOLASSES PROCESSING PLANT[a]

Process stream	kg/day	Process stream	kg/day
Molasses	27,216	Reflux, second effect	26,410
Triple super phosphate	91	Distillate, second effect	75,459
Acetic acid (80%)	449	Fresh syrup	23,814
Water	157,173	Solids recycle	2330
Slurry steam	8225	Total syrup	26,144
Raw mash	193,154	Steam to syrup	3151
Cooker steam	38,474	Steam to stripper	82,854
Cooker mash	231,628	Residue	27,078
Cooler vapor	44,580	Vapor	81,920
Cooled mash	187,048	Water	23,863
Condenser water	767,491	Steam to water	789
Cooler water	212,725	Scrubbed vapor	80,987
Ammonium hydroxide (26%)	614	Scrubbed product	25,586
Corn oil	37	50% Sodium hydroxide	436
Gas	6804	Scrubber product A	26,022
Fermented mash	180,896	Steam to evaporator 1	24,181
Ethanol column vapor	39,917	Vapor from evaporator	23,817
Condensate, feed preheater	18,643	Solids recycle 1	2205
Condensate, final condenser	17,645	Vapor from column 1	21,189
Reflux	36,288	Reflux 1	6887
High wine	3629	Distillate 1	14,302
Liquid below feed	226,882	Water to condenser 1	160,571
Vapor below feed	49,615	Steam to reboiler	1176
Condenser water	240,757	Crude Diol	9515
Steam	53,207	Vapor from column 2	6554
Slops	177,267	Reflux 2	1862
Second stage feed	101,808	Distillate 2	4694
Syrup	23,814	Water to condenser 2	326,592
Scrubbed vapor	80,987	Steam to calandria	7514
Fresh steam	30,780	Impure diol	4821
Total steam	111,767	50% Sodium hydroxide	16
Condenser water	898,309	Solids recycle 3	125
Vapor, first effect	105,292	Steam to still	1304
Reflux, first effect	27,298	Water to condenser 3	20,730
Distillate, first effect	77,994	Final product	4712
Vapor, second effect	101,869	Solids recycle	2330

[a] Source: Wheat (1953c).

Wheat et al. (1948). By employing the amylolytic bacterium B. *polymyxa* in the process, the need for substrate saccharification was eliminated, and the more economically attractive levo isomer of the diol was generated. A flowsheet of the operation is presented in Fig. 14.

TABLE XIV

SUMMARY OF INITIAL COST OF A 27,200-kg/day MOLASSES PROCESSING PLANT[a]

Item	Cost (1952 $)	% of total cost	Chemical Engineering cost indexes 1984/1952	Ratio	Revised cost (1984 $)	% of total cost	Typical % of total cost
Physical costs							
Land and improvements	51,180	1.84	—	—	225,600	1.9	3–7
Buildings	126,550	4.55	300.3/88.5	3.39	429,400	3.6	3–18
Process equipment	1,133,620	40.79	344.0/77.8	4.42	5,012,400	42.3	21–54
Process piping	380,550	13.69	381.2/73.8	5.17	1,965,700	16.6	3–20
Instruments	109,340	3.93	319.1/80.0	3.99	436,100	3.7	2–8
Electrical installations	81,040	2.92	248.0/79.3	3.13	253,400	2.1	2–10
Service facilities	205,370	7.39	—	—	815,300	6.9	8–20
Total	2,087,650	75.11	—	—	9,137,900	77.1	—
Other costs							
Contingency	196,890	7.09	—	—	772,300	6.5	5–15
Insurance and taxes	59,060	2.12	—	—	220,600	1.9	—
Engineering	435,830	15.68	336.3/84.8	3.97	1,728,400	14.6	4–21
Total	691,780	24.89			2,721,300	22.9	—
Total Cost	**2,779,430**	**100.00**	322.7/81.3	3.97	**11,859,200**	**100.0**	—

[a] Sources: original process cost estimation, Wheat (1953c); cost indexes, Chemical Engineering journal; typical percentage of total cost data, Peters and Timmerhaus (1980).

TABLE XV

Summary of Daily Production Costs for a 27,200-kg/day Molasses Processing Plant[a]

Item	Cost estimate (1952)		Cost estimate (1984)	
	1952 $/day	% Total production	1984 $/day	% Total production
Fixed costs				
Management	106.67	4.01	471.72	3.92
Insurance	34.79	1.31	324.91	2.70
Taxes	139.18	5.24	649.82	5.40
Depreciation	466.03	17.53	3284.39	27.27
Total	746.67	28.09	4730.84	39.28
Raw materials				
Molasses	855.90	32.20	1524.00	12.65
Acetic acid	120.88	4.55	247.55	2.06
Ammonium hydroxide	75.15	2.83	154.42	1.28
Corn oil	33.69	1.27	22.55	0.19
Triple superphosphate	7.76	0.29	16.00	0.13
Total	1093.38	41.14	1964.52	16.31
Labor	320.48	12.06	1973.42	16.38
Energy				
Steam	330.00	12.42	1500.00	12.45
Water	52.00	1.97	390.00	3.24
Electricity	31.50	1.17	283.50	2.35
Total	413.50	15.56	2173.50	18.05
Miscellaneous	15.00	0.56	64.98	0.54
Maintenance	68.77	2.59	1137.18	9.44
Total operating cost	**1911.13**	**71.91**	**7313.60**	**60.72**
Total production cost	**2657.80**	**100.00**	**12,044.44**	**100.00**

[a]Sources: original cost estimate, Wheat (1953c); see test for bases of revised estimates.

The initial mash concentration was approximately 15%. With a wheat starch content of approximately 50% and a fermentation efficiency of 90%, the fermented mash contained 2.32% (w/v) diol and 1.54% ethanol. The mash from two batch fermentations (total volume approximately 4100 liters) was processed during a single run of the recovery unit. The stripping column described in Section VIII,A was

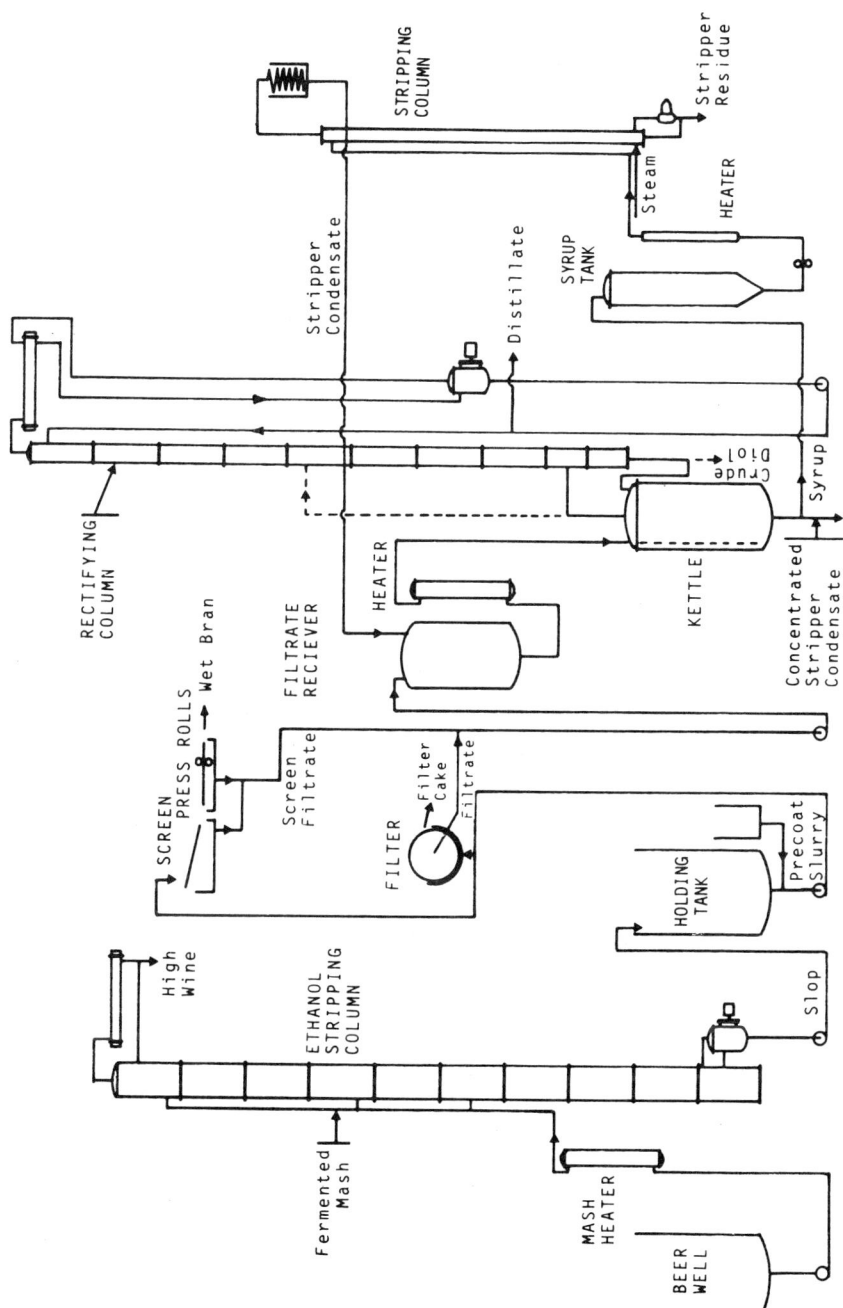

FIG. 14. Flowsheet for pilot-scale production of 2,3-butanediol from wheat (source, Wheat et al., 1948).

employed for ethanol recovery. Solids were removed from the slops by filtration and screening. After an initial charge of approximately 450 liters, the filtrate was heated in a kettle, the vapors from which were fed to a rectifying column. Two options were provided for further purification of the rectified product. Crude diol from the base of the rectifying column was subjected to ethanol precipitation, or syrup from the kettle was steam stripped. In the former method, the crude product was treated with lime before addition of 95% ethanol. After settling was allowed in a storage tank, solids were moved by filtration through cheesecloth. A 95% recovery of the diol was achieved with this method.

In the steam-stripping procedure, syrup from the base of a kettle was preheated before introduction into the top of a 6.1-m stripping column packed with 2.5-cm Raschig rings. Steam was fed into the base of the unit such that the pressure at the top of the column was maintained at 276 kPa (gauge). Diol recovery in the stripping column averaged 97%, producing a condensate which contained 1.5–2.0% (w/v) diol. The condensate also contained 7–10 kg of acid and an equal amount of ammonia for every 100 kg of diol. This solution was then concentrated in the kettle and rectification tower. Evaporation was continued until the volume of the condensate was reduced to approximately 110 liters. The diol was then distilled in two batch operation, the second following the addition of sufficient sodium hydroxide to neutralize acids, saponify esters, liberate ammonia, and precipitate iron and copper.

A process design for the commercial production 2,3-butanediol from 27,200 kg of feed wheat per day was described by Wheat et al. (1948). A flowsheet of the proposal is given in Fig. 15; Table XVI provides mass flow rates and compositions of the various process streams identified. The fermented mash was assumed to have the same composition as that obtained in the pilot plant studies i.e., 2.32% diol and 1.54% ethanol). The pilot ethanol-stripping column described previously was felt to be satisfactory for the commercial operation, and would result in the recovery of 3070 liters of 95% ethanol per batch of mash processed. Screening and pressing of the slops obtained from the base of the stripping column were indicated both for the reduction of product losses and for the improvement of the dried bran composition. The slops were to be concentrated in a double-effect evaporator with forward feed, each effect equipped with a rectifying column for the prevention of diol losses in the vapor streams. Collection of the concentrated syrup in a storage tank enabled intermittent washing of the stripping and scrubbing columns, so that these operations could be performed in an overall continuous manner. Both of these packed columns were specified to be 1.75 m in diameter, while the scrubbing column, at 16.8 m. was taller

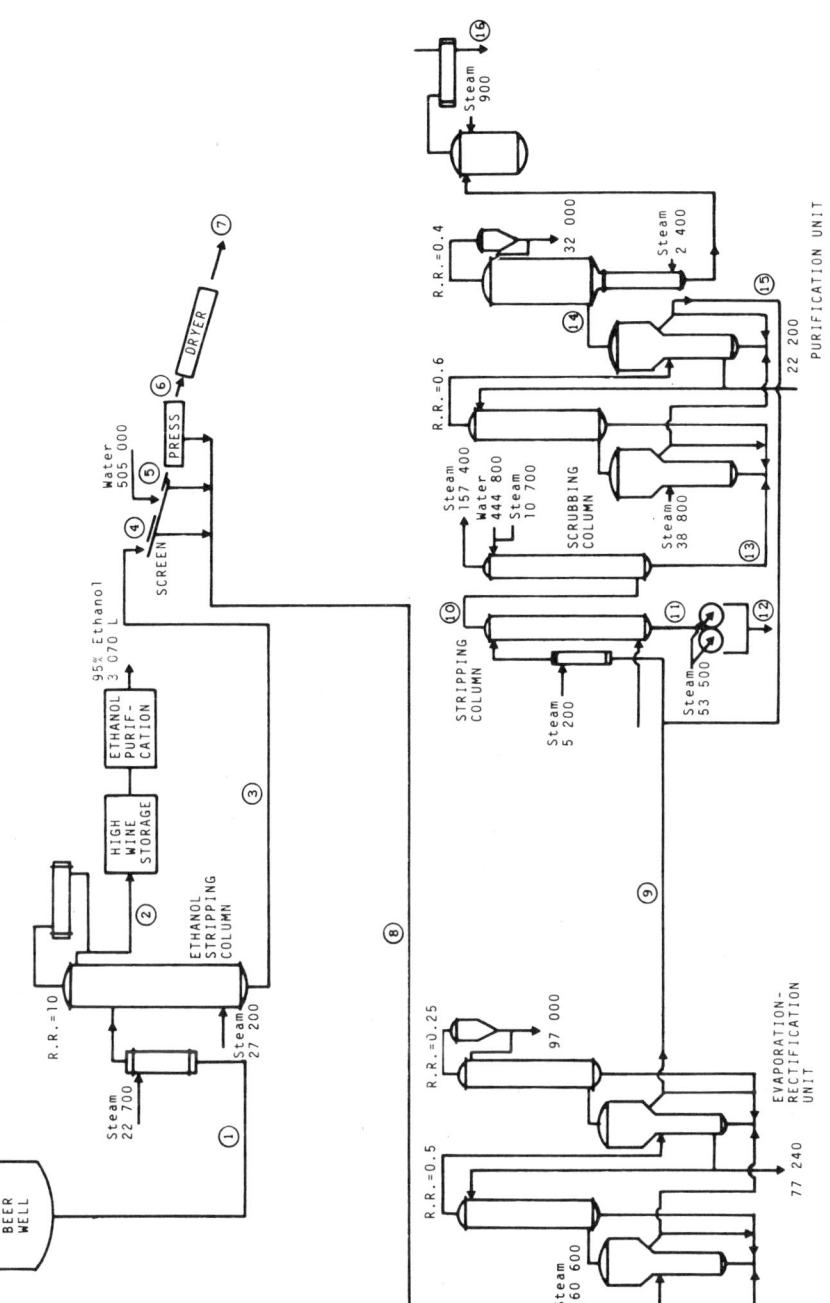

FIG. 15. Flowsheet for commercial operation processing 27 metric tons of wheat per day. Refer to Table XVI for flow rate and composition data for process streams (source, Wheat et al., 1948).

TABLE XVI

MASS FLOW RATES OF PROCESS STREAMS IN A 27,200-kg/day WHEAT PROCESSING PLANT[a]

Process stream	kg/day	Process stream	kg/day
1 Fermented mash		7 Dried bran	
Diol	4042	Diol	16
Ethanol	2694	Solids	1510
Solids	11,249	Water	252
Total	173,953	Total	1778
2 High wine		8 Screen filtrate	
Ethanol	2676	Diol	4017
Diol	4	Solids	9739
Total	3833	Total	215,583
3 Stripping column slop		9 Evaporation–rectification syrup	
Diol	4038	Diol	3946
Insoluble solids	1461	Solids	9739
Soluble solids	9789	Total	41,277
Total solids	11,249	10 Stripper product	
Total	170,120	Diol	3910
4 Screenings		Total	160,435
Diol	394	11 Stripper residue	
Insoluble solids	1461	Diol	36
Soluble solids	971	Solids	9739
Total solids	2431	Total	38,238
Total	17,940	12 Dried solubles	
5 Washed screenings		Diol	32
Diol	97	Water	1706
Insoluble solids	1461	Total	11,476
Soluble solids	238	13 Scrubber product	
Total solids	1699	Diol	3865
Total	17,940	Total	58,150
6 Wet bran		14 Purification vapor	
Diol	20	Diol	3853
Insoluble solids	1461	Total	35,924
Soluble solids	50	15 Solids recycle	
Total solids	1510	Diol	249
Total	4885	Solids	15
		Total	299
		16 Final product	
		Diol	3837
		Total	3914

[a]Source: Wheat et al. (1948).

(stripper height: 11.3 m). Two evaporators were to be used for concentration of the diol to between 97 and 99% (w/v). The first of these units was to be fitted with a rectification column, and a distillation tower was employed to treat the vapor from the second effect. The operation scale proposed would generate 3837 kg of 98% *levo*-2,3-butanediol on a daily basis. By-products of the process would include 3070 liters of 95% ethanol, 1778 kg of dried bran (14.2% water), and 11,476 kg of dried solubles (14.9% water).

The pilot plant studies indicated that the stripper residue (from which the dried solubles are produced) contained 100% of the riboflavin and 50% of the nicotinic acid originally found in the wheat. The protein and mineral content of the dried solubles exceeded that of any residues used commercially as livestock feed at that time. Based on the nitrogen, phosphorus, and potash content of the stripper residue, the material was not felt to have any value as a fertilizer. The dried bran was of even less use for this purpose, and its only feed value was thought to be as roughage.

An economic evaluation of the commercial process (Wheat et al., 1948) indicated that the cost to produce the diol would be 38.1 cents/kg. This estimate included credits of 15.9 cents/liter for the ethanol and 2.95 cents/kg for the dried solids. The initial cost of plant construction was estimated to be $500,000 (1945).

C. BUTANEDIOL PRODUCTION FROM BARLEY

The economic evaluation performed by Wheat et al. (1948) indicated that half of the cost of producing 2,3-butanediol from wheat was due to the substrate. In an attempt, therefore, to develop a process for diol production from less expensive raw materials, Tomkins et al. (1948) developed a pilot-scale system based on the conversion of feed barley by *B. polymyxa*. The most efficient fermentation of the milled feedstock (90% complete after 96 hours) was found to occur at a mash concentration of 12.5%. Based on the pilot plant data, it was estimated that 3550 liters of 98% *levo*-butanediol could be generated in a commercial operation utilizing 27,200 kg of feed barley per day. By-products of the process would include 2770 liters of 95% ethanol, 4640 kg of barley hulls, and 8650 kg of stripper residue (13.5% moisture). Using western Canadian grain prices (1947) of $1.27/bushel of wheat and $0.82/bushel of barley, it was estimated that the cost of producing 2,3-butanediol in a 27,200-kg/day operation would be 43.0 cents/kg from wheat and 40.1 cents/kg from barley. Thus a savings of 7% would result from the substitution of barley for wheat. Recall, however, that the estimated production cost from wheat in the previous section was 38.1 cents/kg.

D. 1,3-Butadiene Production from 2,3-Butanediol

Several methods have been proposed for the conversion of 2,3-butanediol to 1,3-butadiene. Passing diol vapor over thorium dioxide at 350°C under reduced pressure (8.7 kPa) was examined by Winfield (1945). Conversion efficiency was limited, however, to 60% on a single pass. Hale and Miller (1946) studied a technique of diene synthesis in which a mixture of 2,3-butanediol, methyl ethyl ketone, and triethylamine was passed over a mixed catalyst at 235°C. Conversion of the diacetate of 2,3-butanediol in an unpacked tube at 585–595°C was investigated by several groups (Hill and Isaacs, 1938; Schneipp et al., 1945a; Morell, 1945; Morell et al., 1945). The production of diene by passing 2,3-butanediol diacetate over a copper phosphate catalyst in a heated tube was reported by Pines and Ipatieff (1945).

Pyrolysis of diacetate was felt to provide the most effective means of butadiene formation, thus acetylation of the diol represented an essential step in the overall conversion process. Diacetate production in a combined solvent extraction and esterification process as proposed by Othmer et al. (1945) was previously described in Section VII. Pilot-scale production of 2,3-butanediol diacetate by esterification of diol generated from acid-hydrolyzed grain mashes was reported by Grubb (1947). In this method, sulfuric acid was employed in an esterification column to catalyze the formation of diacetate from butanediol and acetic acid. The feed rate to the system was approximately 70 kg of diol per hour. A design was proposed for a commercial-scale operation capable of forming approximately 100,000 kg of 2,3-butanediol diacetate per day.

A pilot plant pyrolysis unit for the formation of 1,3-butadiene from 2,3-butanediol diacetate was described by Schneipp et al. (1945b). At 595°C, the pyrolysis resulted in a single-pass conversion efficiency of 83%. Repyrolysis of the reaction by-products resulted in an 88% yield. The 1,3-butadiene obtained following scrubbing and drying was 99–100% pure and was shown to be suitable for synthetic rubber manufacture. Acetic acid recovery was found to be 99%.

IX. Economics

Humphrey (1975) has stated that biomass resources can be used for the generation of four main product classes: food, feed, chemicals, and energy. Formation of single-cell protein for food or feed is not the primary concern of this review. Proteinaceous by-products have, however, been shown to improve the economic feasibility of the processes discussed in Section 8. Whereas biomass supplies are inadequate for complete replacement of petrochemically derived fuels (Palsson et al., 1981), a role in reducing the dependence on foreign oil has been envisaged. Working with a model of the United States chemicals industry developed by linear programming, Culberson and Donaldson (1982) demonstrated that biomass could be utilized to completely satisfy the demand for the 25 key compounds that account for 90% of United States organic chemical production. The commercial practicality of a biomass-based chemical feedstock industry is enhanced by the fact that the chemical value of a reactive compound is generally triple that of its fuel value (Palsson et al., 1981). Under the conditions of the model developed by Culberson and Donaldson, however, only 35% of the production costs could be recovered from the sale of the compounds produced.

Two additional major difficulties are foreseen in market acceptance of fermentation products (Palsson et al., 1981). The outlets for high-value chemicals such as alcohols are limited, and thus transformation to intermediates (of lower value) would be required in order to achieve major market penetration. In addition, the highly interactive nature of the petrochemical industry would result in the creation of severe imbalances if a single compound was introduced on a large scale. In an effort to minimize the impact upon the existing industry, simultaneous introduction of a mixture of products was recommended.

At current prices, demand for petrochemicals in the United States is well below plant capacity (Chem. Eng. News, 1984), and thus production of fermentation chemicals must be highly competitive. In Canada, however, the high cost of feedstocks in petrochemical production is seen to be a major problem, so that efficient bioconversion of waste materials may result in economic feasibility.

As the primary feedstocks of the petrochemical industry, the yearly demand for short-chain olefins and aromatics (e.g., toluene and xylene) typically exceeds 4.5×10^8 kg (Palsson et al., 1981). Dehydration of alcohols results in the formation of short-chain olefins, for example, the production of butadiene from butanediol. With gasification providing the most efficient means of generating one-carbon compounds, the most promising fermentation products would therefore be those in the two- to four-carbon range. Included in this class are ethanol, acetic acid, acetone, isopropanol, n-butanol, and 2,3-butanediol. Recent United States prices and production statistics (where available) for these compounds are listed in Table XVII.

A detailed economic assessment of commercial diol production has not been published since the work of Wheat in 1953. In an effort to

TABLE XVII

CURRENT SELLING PRICES AND PRODUCTION STATISTICS (UNITED SSTATES) FOR SELECTED ORGANIC CHEMICALS[a]

Chemical	Cost (1985 U.S. $)	Production (10^9 kg)
Acetic acid (fermentation, tank)	0.55/kg	1.27
Acetone (tank)	0.51/kg	0.84
1,3-Butadiene (tank)	0.66/kg	1.18
1,3-Butanediol (tank)	1.59/kg	—
1,4-Butanediol (tank)	1.76/kg	—
n-Butanol (synthesis, fermentation, tank)	0.79/kg	—
Ethanol (fermentation, tank)	0.39–0.45/liter	—
Ethanol (synthesis, 190 proof, tank)	0.48–0.50/liter	—
Ethanol (absolute, 200 proof, tank)	0.51–0.53/liter	—
Ethylene	0.49–0.51/kg	14.60
Diacetyl (flavor grade, drum)	20.39–33.07/kg	—
Isopropanol (anhydrous, 99%, tank)	0.53/liter	—
Methanol (synthesis, tank)	0.14–0.19/liter	3.86
Methyl ethyl ketone (tank)	0.79/kg	—
Styrene (monomer, 99.6% min)	0.57–0.79/kg	3.49
Sulfuric acid (100%, tank)	61.00–95.90/ton	39.52

[a] Source: prices, *Chemical Marketing Reporter*, Mar. 18, 1985; production data, *Chemical Engineering News*, Dec. 17, 1984.

examine the current feasibility of such a process, the cost data of Wheat's study (discussed in Section VIII,A) have been converted here to 1984 United States dollars. (Refer to Tables XIV and XV for the updated cost data.) Revision of the capital and production costs associated with the synthesis of diol from molasses was performed according to guidelines presented in Peters and Timmerhaus (1980) as follows.

A. CAPITAL COSTS (TABLE XIV)

1. The revised cost of plant buildings was calculated from the appropriate plant cost indexes located in the *Chemical Engineering* journal.

2. Equipment cost indexes in this journal were used to update costs of process equipment, piping, instruments, and electrical facilities.

3. The engineering cost determined in the original study was revised with the use of the appropriate plant cost index.

4. The cost of land and improvements was estimated to be 4.5% of the revised purchased equipment cost (factor used by Wheat for determination of the 1952 cost).

5. The cost of service facilities was estimated at 6.9% of fixed capital investment (FCI). This proportion is lower than that recommended in Peters and Timmerhaus, but is believed to be reasonable because utilities for the process are to be purchased rather than generated at the site. The ratio used is approximately equal to that employed in the original cost estimate.

6. Contingency estimated to be 6.5% of FCI.

7. Insurance and taxes estimated to be 1.9% of FCI.

The total initial cost (in 1984 United States dollars) is thus estimated to be $11,859,200. The contribution of each of the capital costs as a percentage of this "fixed capital investment" is indicated in Table XIV. Note that a reasonable agreement is obtained between these values and the typical cost percentage for chemical process industries as given in Peters and Timmerhaus (1980). As a final check, the revised FCI was estimated directly from the original value through the use of the plant cost index in the *Chemical Engineering* indexes:

$$\$2,779,430(322.7/81.3) = \$11,032,252$$

The value obtained in this manner is within 7% of the detailed estimate.

B. Production Costs (Table XV)

1. Management cost estimated at 25% of labor cost (see step 6).
2. Insurance calculated as 1% of FCI/acre.
3. Taxes calculated as 2% of FCI.
4. Depreciation estimated as 10% of FCI (for equipment) plus 3% of initial building cost.
5. Raw materials costs: prices for acetic acid, ammonium hydroxide, corn oil, and triple superphosphate obtained from Dec. 31, 1984 issue of *Chemical Marketing Reporter*. Cost of molasses based on sucrose content (52 wt.%) and average world price for sugar (10.76 cents/kg from *Wall Street Journal*, Dec. 21, 1984.
6. Labor rates obtained from *U.S. Survey of Current Business*: average hourly wage for operators in chemical process industries for Dec. 1984, $11.38; wage for supervisors estimated at 1.2 times operator wage, $13.66; labor cost based on original estimate of operator requirement of 135 man-hours/day, supervisor requirement of 32 man-hours/day.
7. Energy costs: steam, 272,200 kg/day at $5.51/1000 kg; process water, 2,460,500 liters/day at $0.16/1000 liter; electricity, 6,300 kw-hr/day at $0.045/kw-hr.

8. Allowance for miscellaneous expenses based on original estimate of 0.2% of FCI/day.

9. Estimate of maintenance cost based on arithmetic average of recommended value (6% of FCI/acre; Peters and Timmerhause), and original estimate (0.9% of FCI/acre).

Note that in the original economic evaluation of the process (Wheat, 1953c) the initial capital investment ($2,780,000) was found to greatly exceed that of a similar plant (Wheat et al., 1948) designed for the conversion of wheat to diol ($865,000 when converted to 1952 prices). The reason for the observed tripling of capital cost was attributed to the extensive use of stainless-steel equipment in the molasses plant. The selling price of the diol was, however, believed to be below that from the wheat if substitution of alternative materials of construction for major pieces of equipment in the molasses plant was undertaken. A similar reduction in the estimated 1984 capital cost may also be possible.

Based on the data presented in Table XV, the 1984 production cost for the diol would be $12,044.44/4712.4 kg, or $2.56/kg. This is approximately 1.6 times the current selling price for 1,3-butanediol (Table XVII). Credits obtained from the sale of process by-products should help to significantly reduce the final price of the diol. At its current market value of approximately 42 cents/liter, the income generated from the sale of the 3343 liter of 95% ethanol produced daily would total $1400. Material balances for the operation (Table XIII) indicate a daily production of 27,078 kg of stripper residue. Based on studies of wheat conversion in a similar process design (Wheat et al., 1948; Section VIII,B), the anticipated yield of dried solubles from this residue would be approximately 8200 kg/day. Analysis of the residue obtained from the wheat operation indicated that while its nitrogen, phosphorus, and potash content was not sufficient to enable successful marketing as fertilizer, its protein and mineral composition was superior to that of any commercially available livestock feed. The nutritionally favorable composition of beet molasses (Nathan, 1978) should therefore indicate that this material could be introduced as feed at a relatively high value. At an intermediate price of $120/metric ton, the revenue from the sale of the dried solubles would total $972/day. The income from the ethanol and dried solubles would thus reduce the daily production costs to approximately $9670. The break-even price for the diol would as a result be reduced to $2.05/kg.

If it is assumed that the plant could operate at full capacity for 260 days/year, the total annual production of diol would be 1,225,120 kg. If a target return on the initial capital investment of 10% was set, then the additional increase in the price of the diol would be 0.10($11,859,200)/(1,225,120 kg/year) or $0.97/kg. The final selling price of the product would thus be $3.02/kg. Reduction in the capital cost through the use of less expensive materials of construction for process equipment, utilization of a less expensive sugar feedstock, and the application of more efficient process technologies could conceivably result in the generation of a competitive product.

Researchers at New York's Rensselaer Polytechnic Institute have recently stated their belief (*Eur. Chem. News*, 1984) that continuous production of *levo*-2,3-butanediol from corn by *B. polymyxa* can be performed at a 1.8×10^8-kg scale at a cost below that of conventional 1,4-butanediol processes (less than $1.76/kg). The fermentation-based diol could thus replace 1,4-butanediol in the synthesis of both specialty and commodity grades of polyesters and polyurethanes. Reduction of production costs by a factor of 3 was believed possible through the use of cellulose-derived glucose in place of corn.

X. Conclusions

The formation of 2,3-butanediol by bacterial species, first noted in the late nineteenth century and developed to pilot plant scale in the 1940s, continues to be of interest today by virtue of the diverse applications of this compound. Two microbial species have demonstrated a potential for diol production on a commercial scale. *Klebsiella pneumoniae*, with broad substrate and environmental adaptability, is the most thoroughly investigated organism. The ability to utilize starchy feedstocks is the main advantage of *Bacillus polymya*. Utilization of "waste" cellulosic substrates is generally recommended for improvement of process economics. Efficient conversion of hemicellulosic carbohydrates is thus essential. The optimal conditions for bioconversion of glucose are clearly distinct from those for xylose. This is especially true in terms of culture pH and aeration, and to a lesser extent in temperature and nutrient supplementation. The most effective process designs may therefore permit separate conditions for the consumption of these sugars. In a batch operation, this could be accomplished through a mid-run adjustment of key environmental parameters. In continuous production, separate reactors could be operated in a series with residence times and conditions established for optimal consumption of the glucose and xylose. Recovery of butanediol from fermented broths is especially difficult due to the physical and chemical properties of the compound. Solvent extraction and steam stripping appear to be the most effective alternatives at present. Incentives to seek renewable alternatives to petroleum-based fuels and chemicals in combination with advances in microbial process efficiency may yet result in an economically viable system for the production of butanediol from biomass resources.

REFERENCES

Adams, G. A., and Leslie, J. D. (1946). Can. J. Res. **24F**, 107–116.
Anastassiadis, P. A., and Wheat, J. A. (1953). Can. J. Technol. **31**, 1–8.
Bahadur, K., and Ranganayaki, S. (1960). Indian J. Appl. Chem. **23**, 3–8.
Bailey, J., and Ollis, D. (1977). "Biochemical Engineering Fundamentals," McGraw-Hill, p. 411–426. New York.
Birkinshaw, J. H., Charles, J., and Clutterbuck, P. (1931). Biochem. J. **25**, 1522.
Blackwood, A. C., Neish, A. C., Brown, W. E., and Ledingham, G. A. (1947). Can. J. Res. **25B**, 56–64.
Blackwood, A. C., Wheat, J. A., Leslie, J. D., Ledingham, G. A., and Simpson, F. J. (1949). Can. J. Res. **27F**, 199–210.
Blanco, R. M., Gillman, L., Cochet, N., Deschamps, A., and Lebeault, J. M. (1984). Biotechnol. 3rd, Munich, 10–14 Sept.
Blom, R. H., Reed, D. L., Efron, A., and Mustakas, G. C. (1945). Ind. Eng. Chem. **37**, 865–870.
Buchanan, R. E., ed. (1974). "Bergey's Manual of Determinative Bacteriology," 8th Ed. William Wilkins, Baltimore.
Bu'Lock, J. D. (1975). Octagon Pap. **2**, 5–19.
Chem. Eng. **58**, 164 (1951).
Chem. Eng. **April 28**, 117–118 (1975).
Chem. Eng. **April 29**, 75–76 (1985).
Chem. Eng. News **30**, 5110 (1952).
Chem Eng. News **62**, 32–60 (1984).
Chua, J. W., Erarslan, A., Kinoshita, S., and Taguchi, H. (1980). J. Ferment. Technol. **58**, 123–127.
Clendenning, K. A. (1946). Can. J. Res. **24F**, 249–271.
Clendenning, K. A., and Wright, D. E. (1946). Can. J. Res. **24F**, 287–299.
Cooper, J. L. (1976). Biotechnol. Bioeng. Symp. **6**, 251–271.
Culberson, O. L., and Donaldson, T. L. (1982). Biotechnol. Bioeng. Symp. **12**, 291–296.
Detroy, R. W., and Hesseltine, C. W. (1978). Proc. Biochem. **13**, 2.
Donker, H. J. L. (1926). Ph.D. thesis, Delft.
Esener, A. A., Roels, J. A., and Kossen, N. W. F. (1981a). Biotechnol. Bioeng. **23**, 1401.
Esener, A. A., Bol, G. A., Kossen, N. W. F., and Roels, J. A. (1981b). Proc. Int. Ferment. Symp. 6th **1**, 339–344.
Esener, A. A., Roels, J. A., and Kossen, N. W. F. (1983). Biotechnol. Bioeng. **25**, 2093–2098.
Eur. Chem. News **June 4** 19 (1984).
FAO (1974). FAO Bibliogr. List **23**, (27-283-74) 12.
Fields, M. L., and Richmond, B. (1967). Appl. Microbiol. **15**, 1313–1315.
Flickinger, M. C. (1980). Biotechnol. Bioeng. Symp. **22** (Supl. 1), 27–48.
Forage, A. J., and Righelato, R. C. (1979). In "Microbial Biomass" (A. H. Rose, ed), p. 289. Academic Press, New York.

Fratkin, S. B., and Adams, G. A. (1946). *Can. J. Res.* **24F**, 29–38.
Freeman, G. G. (1947). *Biochem. J.* **41**, 389–398.
Freeman, G. G., and Morrison, R. I. (1947). *J. Soc. Chem. Ind.* **66**, 216–221.
Fulmer, E. I., Christensen, L. M., and Kendall, A. R. (1933). *Ind. Eng. Chem.* **25**, 798–900.
Gottshalk, G. (1979). "Bacterial Metabolism." Springer-Verlag, New York.
Grant, C. L., and Pramer, D. (1962). *J. Bacteriol.* **84**, 869–870.
Griffith, W. L., Hulseman, R. A., Googin, J. M., and Compere, A. L. (1982). *AIChE Winter meet., Orlando, Florida. Feb. 28–Mar. 3.*
Grubb, H. H. (1947). *Chem. Eng. Prog.* **43**, 437–448.
Hale, R., and Miller, H. (1946). U.S. Patent 2,400,409.
Happold, F., and Spencer, C. (1952). *Biochim. Biophys. Acta* **8**, 543.
Harden, A., and Norris, D. (1912). *Proc. R. Soc. (London) Ser. B***85**, 415–417.
Harden, A., and Walpole, G. S. (1906). *Proc. R. Soc. (London) Ser. B***77**, 399–405.
Harrison, D. E. F., and Pirt, S. J. (1967). *J. Gen. Microbiol.* **46**, 193–211.
Hill, R., and Isaacs, E. (1938). British Patent 483,989.
Hohn-Bentz, H., and Radler, F. (1978). *Arch. Microbiol.* **116**, 197–203.
Horecker, B. L. (1962). "Pentose Metabolism in Bacteria." Wiley, New York.
Humphrey, A. E. (1975). *Biotechnol. Bioeng. Symp.* **5**, 49–65.
Jansen, N., and Tsao, G. (1983). *Avd. Biochem. Eng.* **27**, 85–99.
Jansen, N. B. (1982). PhD thesis, Purdue University.
Jansen, N. B., Flickinger, M. C., and Tsao, G. T. (1984). *Biotechnol. Bioeng.* **26**, 362–369.
Johansen, L., Larsen, S. H., and Stormer, F. C. (1973). *Eur. J. Biochem.* **34**, 97–99.
Johansen, L., Bryn, K., and Stormer, F. (1975). *J. Bacteriol.* **123**, 1124–1130.
Katznelson, H. (1944a). *Can. J. Res.* **22C**, 235–240.
Katznelson, H. (1944b). *Can. J. Res.* **22C**, 241–250.
Kinn, R. K. (1967). *In* "Biochemical and Biological Engineering Science" N. Blakebrough, (ed) p. 81. Academic Press, New York.
Kluyver, A. J., and Donker, H. J. L. (1925). *Akad. Wetenschappen Amsterdam* **28**, 314–317.
Kolfenbach, J., Kooi, E., Fulmer, E., and Underkofler, L. (1944). *Ind. Eng. Chem. Anal. Ed.* **16**, 473–474.
Kosaric, N., Weiczorek, A., Cosentino, G. P., Magee, R. J., and Prenosil, J. E. (1983). *Biotechnology* **3**, 257.
Kosikowski, F. V. (1979). *J. Dairy Sci.* **62**, 1149–1160.
Larsen, S. H., and Stormer, F. C. (1973). *Eur. J. Biochem.* **34**, 100–106.
Laube, V. M., Groleau, D., and Martin, S. M. (1984a). *Biotechnol. Lett.* **6**, 257–262.
Laube, V. M., Groleau, D., and Martin, S. M. (1984b). *Biotechnol. Lett.* **6**, 535–540.
Ledingham, G. A., and Neish, A. C. (1954). *In* "Industrial Fermentations" (L. A. Underkofler and R. J. Hickey eds), pp. 27–93. Vol 2. Chemical Publ., New York.
Ledingham, G. A., and Stanier, R. Y. (1944). *J. Bacteriol.* **47**, 443.
Ledingham, G. A., Adams, G. A., and Stanier, R. Y. (1945). *Can. J. Res.* **23F**, 48.
Lee, H. K., and Maddow, I. S. (1984). *Biotechnol. Lett.* **6**, 815–818.
Liebmann, J. (1945). *Oil Soap* **22**, 31–34.
Loken, J. P., and Stormer, F. C. (1970). *Eur. J. Biochem.* **14**, 133–137.
Long, S. K., and Patrick, R. (1960). *Proc. Florida State Hortic. Soc.* **73**, 241–246.
Long, S. K., and Patrick, R. (1961). *Appl. Microbiol.* **9**, 244–248.

Long, S. K., and Patrick, R. (1963). The present status of the 2,3-butylene glycol fermentation. *Adv. Appl. Microbiol.* **5,** 135–156.
Long, S. K., and Patrick, R. (1965). *Appl. Microbiol.* **13,** 973–976.
McCall, K. B. (1943). M. A. thesis, University of Nebraska, Lincoln.
McCall, K. B., and Georgi, C. E. (1954). *Appl. Microbiol.* **2,** 355–359.
Mahmoud, S. A. Z., Taha, S. M., El-Sadek, G. M., and Ali, S. A. (1975). *Rev. Microbiol.* (S. Paulo) **6,** 77–80.
Malthe-Sorenssen, D., and Stormer, F. C. (1970). *Eur. J. Biochem.* **14,** 127–132.
Mes-Hartree, M., and Saddler, J. N. (1983). *Biotechnol. Lett.* **5,** 531–536.
Mickelson, M., and Werkman, C. (1938). *J. Bacteriol.* **36,** 67–76.
Mickelson, M. N., and Werkman, C. H. (1939). *J. Bacteriol.* **37,** 619–628.
Morell, S. A. (1945). U.S. Patent 2,372,221.
Morell, S. A., and Auernheimer, A. H. (1944). *J. Am. Chem. Soc.* **66,** 792–796.
Morell, S. A., Geller, H., and Lathrop, E. (1945). *Ind. Eng. Chem.* **37,** 877–884.
Mortlock, R. P., and Wood, W. A. (1964). *J. Bacteriol.* **88,** 845–849.
Murphy, D., and Stranks, D. W. (1951). *Can. J. Technol.* **29,** 413–420.
Murphy, D., Stranks, D. W., and Harmsen, G. W. (1951a). *Can. J. Technol.* **29,** 131–143.
Murphy, D., Watson, R. W., Muirhead, D. R., and Barnwell, J. L. (1951b). *Can. J. Technol.* **29,** 375–381.
Nathan, R. A. (1978). Fuels from sugar crops. USDOE Tech. Information Center, Oak Ridge, Tennessee.
Neish, A. C. (1945). *Can. J. Res.* **23B,** 10–16.
Neish, A. C. (1950). *Can. J. Res.* **28B,** 660–661.
Neish, A. C., and Ledingham, G. A. (1949). *Can. J. Res.* **27B,** 694.
Neish, A. C., Blackwood, A. C., Robertson, F. M., and Ledingham, G. A. (1947). *Can. J. Res.* **25B,** 65–69.
Neish, A. C., Blackwood, A. C., Robertson, F. M., and Ledingham, G. A. (1948). *Can. J. Res.* **26B,** 335–342.
Olson, B. H., and Johnson, M. J. (1948). *J. Bacteriol.* **55,** 209–222.
Othmer, D. F., Bergen, W. S., Shlechter, N., and Bruins, P. F. (1945). *Ind. Eng. Chem.* **37,** 890–894.
Owen, W. L. (1950). *Int. Sugar J.* **52,** 120–121.
Palsson, B. O., Fathi-Afshar, S., Rudd, D. F., and Lightfoot, E. N. (1981). *Science* **213,** 513–517.
Pedersen, C. S., and Breed, R. S. (1928). *J. Bacteriol.* **16,** 163.
Perlman, D. (1944). *Ind. Eng. Chem.* **36,** 803–804.
Peters, M. S., and Timmerhaus, K. D. (1980). "Plant Design and Economics for Chemical Engineers," 3rd Ed. pp. 147–221. McGraw-Hill, New York.
Phillips, D. H., and Johnson, M. J. (1961). *Biotechnol. Bioeng.* **3,** 277.
Pines, H., and Ipatieff, V. (1945). U.S. Patent 2,391,508.
Pirt, S. J. (1974). "Principles of Microbe and Cell Cultivation," p. 147. Wiley, New York.
Pirt, S. J., and Callow, D. S. (1958). *J. Appl. Bacteriol.* **21,** 188–205.
Pirt, S. J., and Callow, D. S. (1959). *Select. Sci. Pap. Inst. Super. Sanita* (Rome).
Reynolds, H., and Werkman, C. H. (1937a). *J. Bacteriol.* **33,** 603–614.
Reynolds, H., and Werkman, C. H. (1937b). *Iowa State Coll. J. Sci.* **11,** 373–378.
Roberts, J. L. (1947). *Soil Sci.* **63,** 135–140.
Rosenberg, S. L. (1980). *Enzyme Microb. Technol.* **2,** 185–193.
Sablayrolles, J. M., and Goma, G. (1984). *Biotechnol. Bioeng.* **26,** 148–155.
Saddler, J. N., Yu, E. K. C., Mes-Hartree, M., Levitin, N., and Brownell, H. H. (1983). *AEM* **45,** 153–160.

Salo, D. J., and Henry, J. F. (1979). The report of the Alcohol Fuels Policy Review. DOE/ET-0114/1.
Sankarnarayan, V., Lee, Y. Y., and Chambers, R. P. (1980). *Papermakers Conf.* pp. 175-180.
Sanwall, B. D. (1970). *Bacteriol. Rev.* **34**, 20-39.
Schneipp, L., Dunning, J., Geller, H., Morell, S., and Lathrop, E. (1945a). *Ind. Eng. Chem.* **37**, 884-889.
Schneipp, L., Dunning, J., and Lathrop, E. (1945b). *Ind. Eng. Chem.* **37**, 872-877.
Scott, W. J. (1953). *Aust. J. Biol. Sci.* **6**, 549.
Senkus, M. (1946). *Ind. Eng. Chem.* **38**, 913-916.
Sokatch, J. R. (1969). In "Bacterial Physiology and Metabolism" pp 87-89. Academic Press, New York.
Speckman, R. A., and Collins, E. B. (1982). *Appl. Environ. Microbiol.* **43**, 1216-1218.
Stahly, G. L., and Werkman, C. H. (1942). *Biochem. J.* **36**, 575-581.
Stanier, R. Y., and Adams, G. A. (1944). *Biochem. J.* **38**, 168-171.
Stormer, F. C. (1967). *J. Biol. Chem.* **242**, 1756-1759.
Stormer, F. C. (1968a). *FEBS Lett.* **2**, 36-38.
Stormer, F. C. (1968b). *J. Biol. Chem.* **243**, 3735-3739.
Stormer, F. C. (1977). *Biochem. Biophys. Res. Commun.* **74**, 898-902.
Strecker, H. J., and Harary, I. (1954). *J. Biol. Chem* **211**, 263-270.
Taha, S. M., Mahmoud, S. A. Z., and Ali, S. A. (1971). *U.A.R.J. Microbiol.* **6**, 65-72.
Taylor, G. W., and Juni, E. (1960). *Biochim. Biophys. Acta* **39**, 448.
Tink, R. R. (1950). M.S. thesis, University of Saskatchewan.
Tomkins, R. V., Scott, D. S., and Simpson, F. J. (1948). *Can. J. Res.* **26F**, 497-502.
Tsao, G. T. (1978). Conversion of biomass from agriculture into useful products. Final Report. July 31, 1978. USDOE contract no. EG-77-S-02-4298.
Veeraraghaven, S., Lee, Y. Y., Chambers, R. P., and McCaskey, T. A. (1980). *Enzyme Eng.* **5**, 171-173.
Villet, R., ed. (1981). "Biotechnology for Producing Chemicals from Biomass". Vol. 2. Fermentation Chemicals from Biomass Solar Energy Research Institute. Golden, Colorado. SERI/TR-621-754 (microform).
Vollbrecht, D. (1982). *Eur. J. Appl. Microb. Biotechnol.* **15**, 111-116.
Voloch, M., Ladisch, M. R., Rodwell, V. W., and Tsao, G. T. *Biotechnol. Bioeng.* **25**:173-183.
Weizmann, C., Bergmann, E., Sulzbacher, M., and Pariser, E. (1948). *J. Soc. Chem. Ind.* **67**, 225-227.
Wheat, J. A. (1953a). *Can. J. Technol.* **31**, 73-84.
Wheat, J. A. (1953b). *Can. J. Technol.* **31**, 42-56.
Wheat, J. A. (1953c). *Ind. Eng. Chem.* **45**, 2387-2394.
Wheat, J. A., Leslie, J. D., Tomkins, R. V., Mitton, H. E., Scott, D. S., and Ledingham, G. A. (1948). *Can J. Res.* **26F**, 469-496.
Willetts, A. (1984). *Biotechnol. Lett.* **6**, 263-268.
Winfield, M. (1945). *J. Council Sci. Ind. Res.* **18**, 412.
Yu, E., and Saddler, J. N. (1982a). *Biotechnol. Lett.* **4**, 121-126.
Yu, E., and Saddler, J. N. (1982b). *Appl. Environ. Microbiol.* **44**: 777-784.
Yu, E., and Saddler, J. N. (1983). *Can. Soc. Microbiol.* (Winnepeg) Jn 19-23/83.
Yu, E., Levitin, N., and Saddler, J. N. (1982). *Biotechnol. Lett.* **4**, 741-746.
Yu, E. K. C., Deschatelets, L., and Saddler, J. N. (1984a). *Biotechnol. Lett.* **6**, 327-332.
Yu, E. K. C., Deschatelets, L., and Saddler, J. N. (1984b) *Appl. Microbiol. biotechnol.* **19**, 365-372.
Zemek, J., Augustin, J., Borriss, R., Kuniak, L., Savabova, M., and Pacova, Z. (1981). *Folia Microbiol.* **26**, 403-407.

Microbial Sucrose Phosphorylase: Fermentation Process, Properties, and Biotechnical Applications

ERICK J. VANDAMME, JAN VAN LOO, LIEVE MACHTELINCKX, AND ANDRE DE LAPORTE

Laboratory of General and Industrial Microbiology
State University of Ghent
Coupure Links 653, B-9000 Ghent, Belgium

I. Introduction

Microbial sucrose phosphorylase (disaccharide glycosyltransferase, EC 2.4.1.7) was discovered rather incidentally in 1942 in the Soviet Union. During studies on the dextran sucrase activity of *Leuconostoc mesenteroides* (ATCC 12291), Kagan et al. (1942) observed that sucrose in the presence of inorganic phosphate was phosphorolyzed to glucose-1-phosphate and fructose (Table I). Independently, Doudoroff et al. (1943) demonstrated that dried cell preparations of *Pseudomonas saccharophila* (ATCC 15946) were able to catalyze the same reaction. The reversibility of the reaction was apparent from the fact that glucose-1-phosphate and fructose could be coupled to form sucrose with liberation of equimolar amounts of inorganic phosphate. At that time, no clear picture was as yet available of the biosynthesis of sucrose in plants, despite the extensive use of this sweetener. It was widespread opinion that invertase (EC 3.2.1.26), the enzyme responsible for the hydrolysis of sucrose to glucose and fructose, was equally responsible for sucrose synthesis by a reverse reaction (Oparin and Kursanov, 1931). Several opponents to this theory were not able to prove the true sucrose synthesis mechanism (Lebedew and Dikanowa, 1935; Leonard, 1938). The fact that several enzyme systems in plants and in microorganisms act similarly and that microbes are easily cultured *in vitro* led Doudoroff (1945a,b) to further study microbial sucrose phosphorylase in the hope of gaining more insight into sucrose biosynthesis in general. The true biosynthesis mechanism of sucrose in plants was elucidated in 1953 by Leloir and Cardini (1953); their findings proved that sucrose phosphorylase has no role at all in the formation of sucrose in plants.

II. Producing Strains and Enzyme Assay

As already mentioned, *L. mesenteroides* ATCC 12291 and *P. saccharophila* ATCC 15946 were the first strains discovered to display sucrose phosphorylase activity (Kagan et al., 1942; Doudoroff et al., 1943). In 1949, Doudoroff et al. demonstrated the presence of this

TABLE I

Survey of Disaccharide Phosphorylases

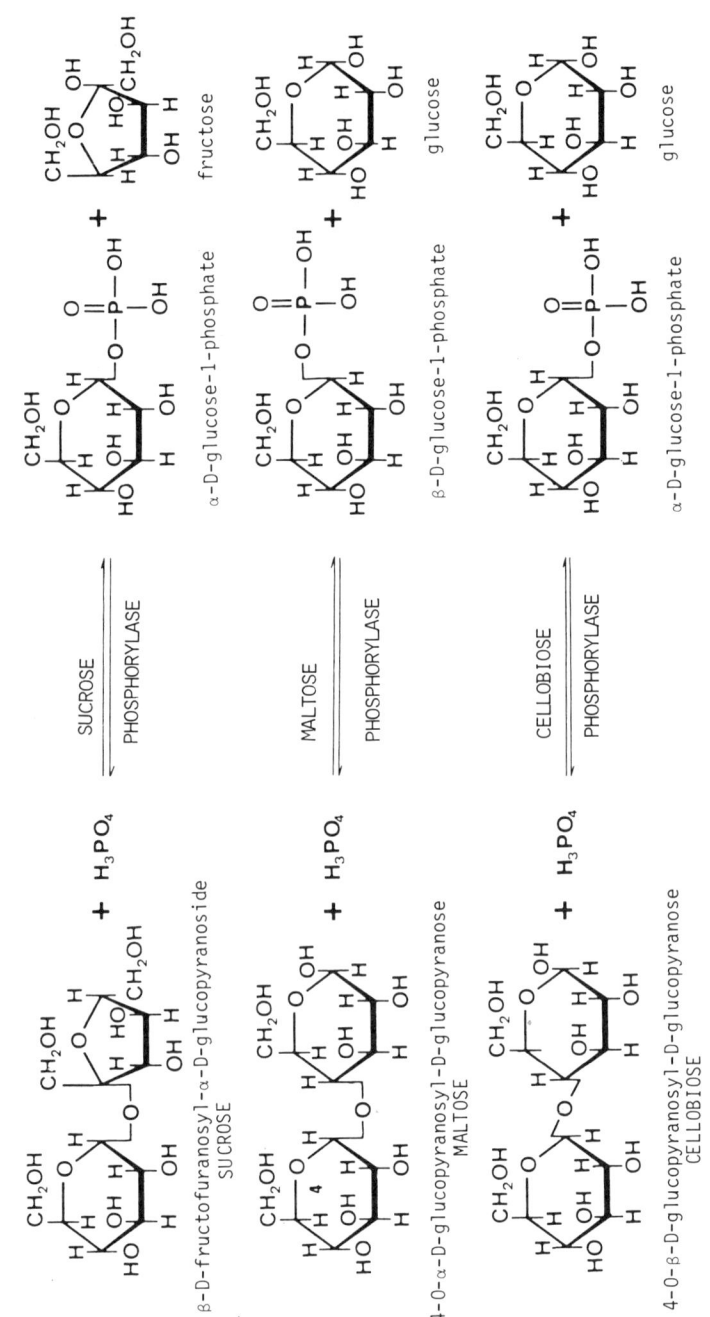

(continued)

TABLE I (Continued)

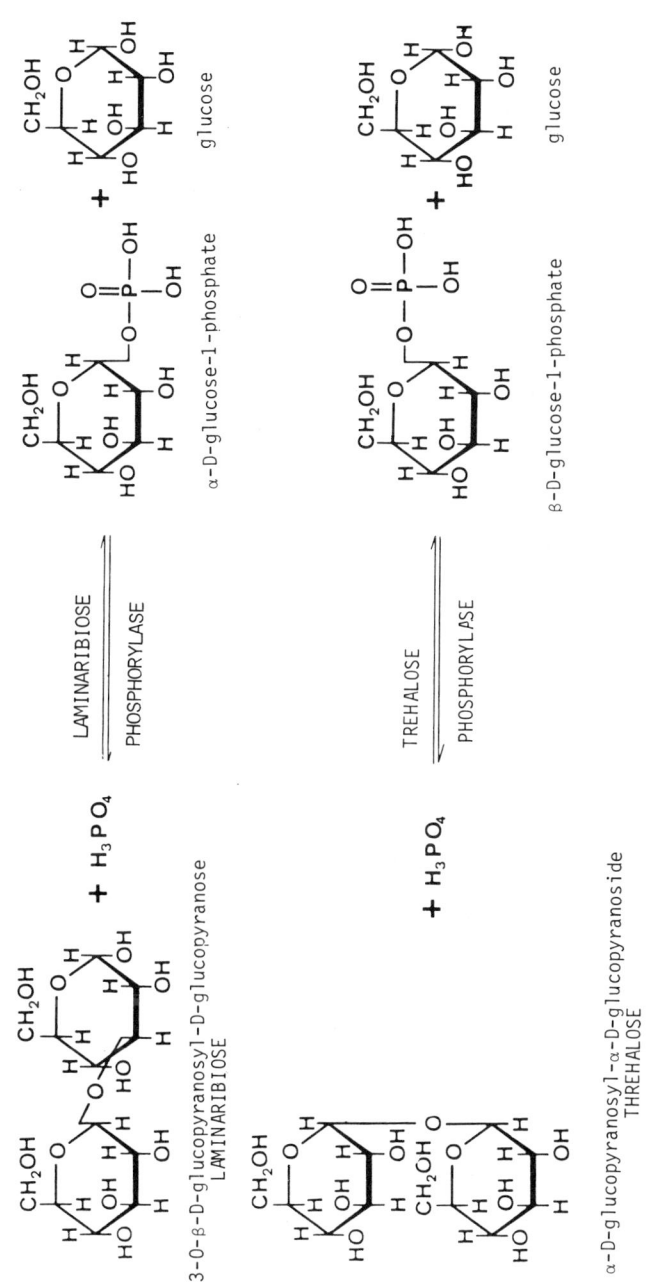

enzyme in *Pseudomonas putrefaciens*. A comparative study of the sucrose phosphorylase activity of these three strains revealed that *L. mesenteroides* displayed the highest activity (Weimberg and Doudoroff, 1954). Until recently, these were the only three types of microorganisms known as producers of the enzyme. In 1980, Kelly and Butler compared the sucrose phosphorylase activity of six strains of *L. mesenteroides*: NRRL B-1051, B-3619, B-1200, B-1430, B-1198, and B-1065; the NRRL B-1065 strain was the best enzyme producer. Also in 1980, Tsai *et al.* reported on a sucrose phosphorylase present in *Clostridium pasteurianum* W5 and Imshenetskii and Kondrateva (1980) mentioned the enzyme activity in the mold *Pullularia pullulans* 1125. The enzyme was also found in *Acetobacter xylinum* (Balasubramaniam and Kannangara, 1982).

The fact that relatively few microorganisms are known to produce sucrose phosphorylase might be related to the fact that so far no simple direct screening test is available for such strains. In addition to sucrose phosphorylase, four other other disaccharide phosphorylases are known today: maltose phosphorylase (EC 2.4.1.8), cellobiose phosphorylase (EC 2.4.1.20), laminaribiose phosphorylase (EC 2.4.1.31), and trehalose phosphorylase (EC 2.4.1.64). The reactions catalyzed by these five disaccharide phosphorylases are presented in Table I. Maltose phosphorylase was first demonstrated in *Neisseria meningitidis* (Fitting and Doudoroff, 1952) and later in *Neisseria perflava* (Selinger and Schramm, 1961) and in *Lactobacillus brevis* (Murao *et al.*, 1985). Cellobiose phosphorylase activity was found in *Clostridium thermocellum* (Sih *et al.*, 1957), *Ruminococcus flavefaciens* (Ayers, 1959), *Cellvibrio gilvus* (ATCC 13127) (Hulcher and King, 1958; Sasaki *et al.*, 1983), and *Cellulomonas* species (*C. uda, C. flavigena, C. cartalyticum,* and *C. fimi*). Laminaribiose phosphorylase is produced by algae (*Euglena gracilis* and *Astasia ocellata* (Marechal and Goldenberg, 1936; Manners and Taylor, 1965). Trehalose phosphorylase is also present in *E. gracilis* (Belocopitow and Marechal, 1970).

Among the five disaccharide phosphorylases discussed herein, sucrose phosphorylase behaves uniquely in that the overall reaction displays no inversion of the configuration of the glucose moiety. A typical reaction mixture for phosphorolysis contains 0.2 M sucrose, 0.1 M KH_2PO_4, 0.2 M $NaHCO_3$, 0.1 M KF (pH 7.3), and enzyme or cells in MOPS buffer. For sucrose synthesis, fructose and glucose-1-phosphate act as substrates. Sucrose and fructose (and eventually the mannitol produced) can be quantified simultaneously by gas chromatographic procedures. (De Laporte, 1982; unpublished results). Sucrose formed or inorganic phosphate liberated can be measured to quantify synthesis activity, whereas the amount of glucose-1-phosphate or fructose is a measure of phosphorolysis. One unit of sucrose phosphorylase will phosphorolyze 1.0 μmol of sucrose to glucose-1-phosphate and fructose per minute at pH 6.7 at 37°C in a reaction mixture containing 0.1 M sucrose and 0.1 M inorganic phosphate.

III. Sucrose Phosphorylase Production

A. MEDIA AND FERMENTATION CONDITIONS

Microbial fermentations related to sucrose phosphorylase production are unsatisfactorily described in the literature. Apart from the strain involved and the medium used, further information concerning process control and parameter changes (pH, temperature, D.O.) during the fermentation cycle is lacking. A survey of fermentation media commonly used is shown in Table II. Doudoroff (1943) cultivated P. saccharophilia in a chemically defined medium (see Table II). Replacing sucrose with other sugars indicated that, apart from sucrose, only raffinose led to enzyme formation. Activity with raffinose as the sole carbon source could be explained by the fact that this trisaccharide is first hydrolyzed by enzyme activity to sucrose and galactose; the thus-liberated sucrose induces sucrose phosphorylase. These facts indicated that the Pseudomonas sucrose phosphorylase is an inducible enzyme. All fermentation media reported are hence based on sucrose as a carbon source. With the Pseudomonas strains, both organic and inorganic nitrogen sources have been used; with the Leuconostoc strains, organic nitrogen (tryptone), yeast extract, and vitamins are essential for growth. On a 10-liter scale, Leuconostoc strains were grown strictly anaerobically (under nitrogen), or microaerophilically (De Laporte, 1982; De Laporte and Vandamme, 1983) with slow agitation to keep a homogeneous suspension; the Pseudomonas strains were aerated constantly with compressed air to ensure efficient agitation (Taylor et al., 1982).

B. GENERAL FERMENTATION PATTERN

The L. mesenteroides ATCC 12291 sucrose phosphorylase fermentation process has recently been studied in detail (De Laporte, 1982; De Laporte et al., 1981, 1982; Vandamme et al., 1984a,b, 1987; De Valck, 1982; Van Loo, 1983). It might serve as an example of a potentially important enzyme fermentation with an anaerobic microorganism as enzyme producer. An optimized medium, based on the composition of De Moss et al. (1951) (see Table II), was devised with sucrose at 20 g/liter as sole

TABLE II

FERMENTATION MEDIA FOR SUCROSE PHOSPHORYLASE PRODUCTION

Medium		Microorganism	Initial pH	Temperature
Sörensen phosphate buffer[a]	0.033 M	P. saccharophila,	6.64	30°C
(KH_2PO_4 + Na_2HPO_4)	0.1%	P. putrefaciens		
NH_4Cl	0.1%			
$MgSO_4$	0.05%			
$FeCl_3$	0.005%			
$CaCl_2$	0.001%			
Yeast extract	0.05%			
Sucrose	0.3%			
Sörensen phosphate buffer[b]	0.033 M	P. saccharophila	6.8	30°C
NH_4Cl	0.1%			
(Fe, NH_4)-citrate	0.005%			
$MgSO_4 \cdot 7H_2O$	0.05%			
Sucrose	0.25%			
$CaCl_2$	1% or 0.005% (with vitamin–mineral solution)			
Sörensen phosphate buffer[c]	0.033 M	P. saccharophila	6.2–6.5	30°C
$(NH_4)_2SO_4$	0.001%			
$MgSO_4$	0.05%			
$NH_4Fe(SO_4)_2$	0.005%			
Sucrose	0.5%			
Tryptone[d]	1%	L. mesenteroides	±7.5	30°C
Yeast extract	1%			
K_2HPO_4	0.5%			
Sucrose	1%			
Vitamin–mineral solution, 0.5% asorbic acid	1%			
Tryptone[e]	4%	L. mesenteroides	±7.5	31°C
Yeast extract	2.5%			
K_2HPO_4	5%			
Sucrose	4%			
Vitamin–mineral solution, 1% asorbic acid	0.025%			

[a]Doudoroff et al. (1943); Weimberg and Doudoroff (1954).
[b]Doudoroff (1955); Taylor et al. (1982); vitamin–mineral solution: thiamin hydrochloride, 0.1%; ascorbic acid, 0.5% or 1%; $MgSO_4 \cdot 7H_2O$, 4%; $FeSO_4 \cdot 7H_2O$, 0.1%; and $MnSO_4 \cdot 4H_2O$, 2%.
[c]Silverstein et al. (1967).
[d]DeMoss et al. (1951).
[e]Chassy and Krichevsky (1972).

carbon source, tryptone and yeast extract (each 10 g/liter), K_2HPO_4 (10 g/liter), and 10 ml/liter of the vitamin–minerals solution (see Table II). Furthermore, it was shown that the L. mesenteroides sucrose phosphorylase is a constitutive enzyme (Van Loo, 1983; Vandamme et al., 1987), unlike the enzyme produced by P. saccharophila (Doudoroff, 1943). This is an important observation in view of culturing the Leuconostoc strain on cheap industrial substrates as carbon source, since the strain is not dependent on sucrose.

Media for sucrose phosphorylase production based on corn steep liquor and molasses have also been tested (De Valck, 1981; De Laporte, 1982). Slower growth and a much lower enzyme activity were obtained. Further research on these aspects is urgently needed in view of large-scale production of this enzyme. Leuconostoc mesenteroides ATCC 12291 does not grow without addition of the vitamin–minerals solution. The initial pH of the fermentation was 7.5; temperature was kept at 30°C and the fermentation was conducted anaerobically with no agitation. A typical 10-liter fermentation pattern is represented in Fig. 1. Sucrose utilization is proportional to growth and parallels the drop in pH, which is mainly caused by accumulation of lactic and acetic acids. Initially, fructose accumulates and concomitantly a side product, identified as mannitol, is formed. When sucrose is exhausted, the accumulated fructose is partially further metabolized and partially converted to mannitol. Within 8 hours after inoculation, the stationary phase is reached and the pH remains at 4.7; the mannitol formed remains intact in the medium. A specific growth rate of $\mu = 0.95\ hr^{-1}$ was calculated; maximal biomass averaged 1.6–1.7 g/liter DCW. A simplified fermentation scheme is as follows (see also Fig. 2):

$$\text{sucrose} + P_i \rightarrow \text{glucose-1-phosphate} + \text{fructose}$$
$$\downarrow \qquad\qquad\qquad\qquad \downarrow$$
$$\text{lactic acid} \qquad\quad \text{mannitol}$$
$$\text{acetic acid}$$
$$\text{ethanol}$$
$$CO_2$$

Values for Y_{ATP} (grams DCW/mole ATP) of 17.9 and 16.3 were calculated for a microaerophilic and for a strictly anaerobic L. mesenteroides ATCC 12291 fermentation, respectively. No invertase activity could be demonstrated. A low level of dextran formation was detected in the microaerophilic fermentation. Sucrose phosphorylase activity rose gradually and reached a peak value at the end of the log phase and then dropped drastically to a low value upon further fermentation. The sucrose phosphorylase activity of the cells thus varies during the

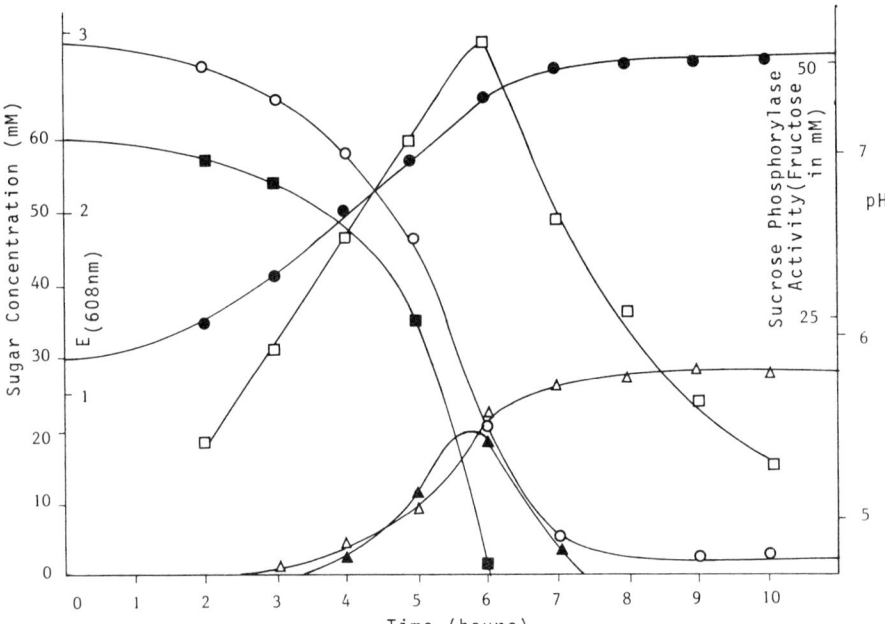

FIG. 1. Leuconostoc mesenteroides fermentation profile on a 10-liter scale: ●, growth; ○, pH; ■, sucrose utilization; △, mannitol; ▲, fructose; □, sucrose phosphorylase activity (De Laporte, 1982).

fermentation cycle such that the moment of cell or enzyme harvest is very critical.

Literature data on peak sucrose phosphorylase activity in other strains are contradictory. *Pseudomonas saccharophila* and *P. putrefaciens* cells were harvested during the late log phase; Taylor et al. (1982) harvested after 2 days of fermentation at room temperature. Doudoroff (1955) and Chassy and Krichevsky (1972) collected *L. mesenteroides* cells immediately upon reaching the stationary phase. Kelly and Butler (1980) found that *L. mesenteroides* NRRL B-1430 biomass formation reached a maximum after 48 hours; specific sucrose phosphorylase production increased during the first 24 hours and then declined rapidly. Daurat-Larroque et al. (1982) harvested *L. mesenteroides* cells grown with 2% glucose as carbon source after 18 hours of fermentation at 30°C. With *P. pullulans*, sucrose phosphorylase activity initially increased during the first 24 hours of fermentation, then dropped to a minimum value after 38 hours, and then increased again up for to 72 hours of fermentation (Imshenetskii and Kondrateva, 1980).

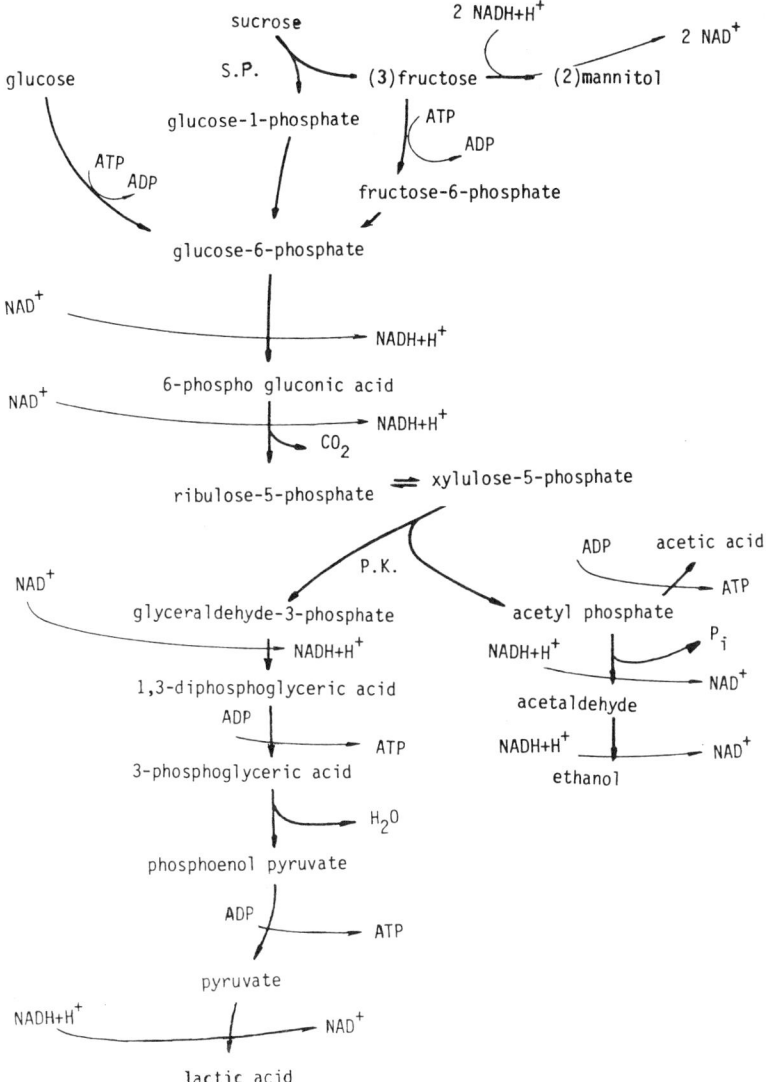

FIG. 2. Metabolism of sucrose, glucose, and fructose by *Leuconostoc mesenteroides*; S.P., sucrose phosphorylase; P.K. phosphoketolase.

C. Effect of Medium pH and of Temperature on Sucrose Phosphorylase Fermentation

In *Leuconostoc* fermentations, sucrose phosphorylase synthesis increases with increasing pH (from 2 U at pH 4.9 to 10 U at pH 7.6) (Fig. 3). Growth rate remains at a comparable level within the pH range 4.9–7.6. The formation of mannitol, here an interesting side product, decreases at high pH values (Van Loo, 1983).

Fermentations were run at different temperatures (Fig. 4) at constant pH 6.5; the highest temperature (38°C) yielded the lowest enzyme level. Mannitol formation increased with higher temperature. Growth rate was highest at 30–35°C. An optimized fermentation run at a controlled pH of 7.3 at 32°C resulted in rapid growth with high levels of enzyme formation. Formation of dextran was observed especially at extremes of pH and temperature in media with high sucrose (100g/liter) substrate concentrations; this resulted in high-viscosity fermentation liquids. Addition of 1 mM maltose could alleviate this problem (Van Loo, 1983); otherwise, dextran-negative mutants could be used.

IV. Properties of Sucrose Phosphorylase

A. Liberation from Cells

Sucrose phosphorylase is an "intracellular" enzyme, though it is not known whether the enzyme is cell wall bound, periplasmic, or truly

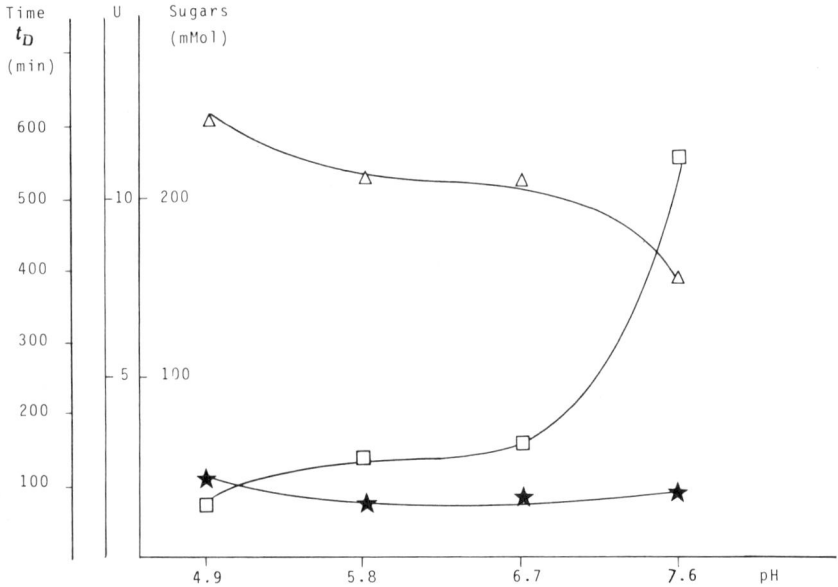

FIG. 3. Effect of pH on mass doubling time t_D (★), enzyme formation (□), and mannitol formation (△) (Van Loo, 1983).

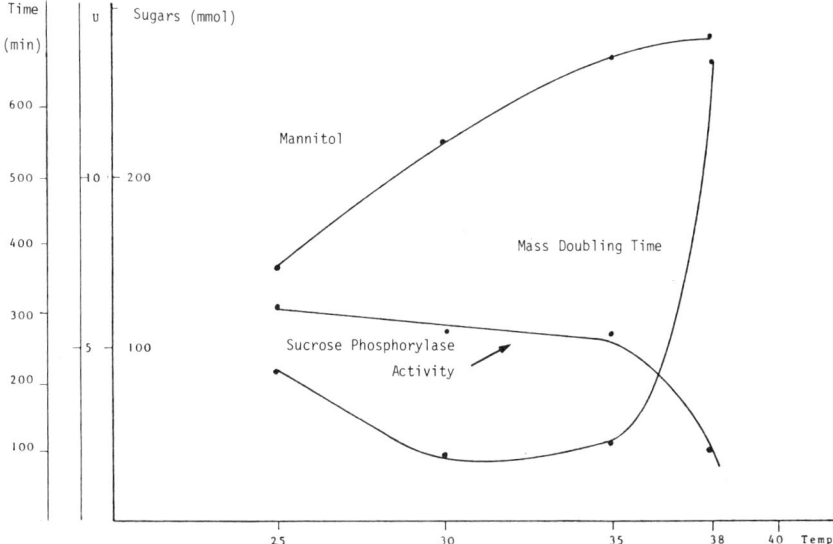

FIG. 4. Temperature effect on t_D, sucrose phosphorylase formation, and mannitol formation.

intracellular (Vandamme et al., 1986). Because several strains producing this enzyme also display phosphatase and invertase (especially *Pseudomonas*) or mannitol dehydrogenase (*Leuconostoc*) activity, cell disruption and enzyme extraction and purification procedures were needed to obtain homogeneous sucrose phosphorylase preparations. Comparison of pure sucrose phosphorylase from different strains revealed differences in molecular weight, substrate specificity, and substrate inhibition pattern.

Initially, Kagan et al. (1942) experimented with intact *L. mesenteroides* cell suspensions, whereas Doudoroff et al. (1943) used P_2O_5-dried *P. saccharophila* cell preparations as the enzyme source. To liberate the enzyme from the cells, Weimberg and Doudoroff (1954) treated wet *P. saccharophila* cells in a mortar with alumina and subsequently extracted the enzyme with 0.039 M phosphate buffer; Abraham and Hassid (1957) mixed P_2O_5-dried cells in a Waring blender; ultrasonic treatment was used by Silverstein et al. (1967), Taylor et al. (1982), Tsai et al. (1980) on *C. pasteurianum* cells as well. *Leuconostoc mesenteroides* cell paste in 0.033 M phosphate buffer (pH 6.8) was passed three times through a French pressure cell (Chassy and Krichevsky, 1972; Daurat-Larroque et al., 1982). De Laporte (1982) compared the sucrose phosphorylase activity of untreated living *L. mesenteroides* cells with that of P_2O_5-dried cells, acetone-dried cells, and lysozyme- or toluene-treated cells. Living cells displayed poor activity and an active mannitol dehydrogenase interfered. P_2O_5-or acetone-dried cell preparations could be easily extracted with phosphate or MOPS buffer to deliver an active enzyme extract free of interfering mannitol dehydrogenase.

B. PURIFICATION PROCEDURES

A comparative study of the three semipurified sucrose phosphorylases from P. saccharophila, P. putrefaciens, and L. mesenteroides by Weimberg and Doudoroff (1954) revealed that different amounts of saturated ammonium sulfate were needed to precipitate the enzymes, indicating differences in enzyme characteristics. The specific activities obtained were 30.6, 13.4, and 142 U, respectively. A 10-fold further increase in purity was obtained for the L. mesenteroides enzyme by Doudoroff (1955) by subsequent acid precipitation with 0.5% acetic acid and by calcium phosphate gel adsorption.

Silverstein et al. (1967) reported a purification scheme (Table III) that yielded homogenous P. saccharophila sucrose phosphorylase. Ultracentrifugation and electrophoresis experiments indicated homogeneity. Sephadex chromatography gave a molecular weight value of 84,000; Tsai et al. (1980) found a molecular weight of 78,000. Upon sodium dodecyl sulfate gel electrophoresis, a molecular weight of 50,000 was found, suggesting the existence of two subunits. A French press cell extract from L. mesenteroides was subjected to ammonium–protamine sulfate treatment followed by dialysis against demineralized water. Further purification on Sephadex G-200 gels yielded a highly active preparation; no molecular weight value is known for this enzyme. An ultrasonic extract of C. pasteurianum was further purified on DEAE cellulose and by gel filtration techniques by Tsai et al. (1980); a molecular weight of 36,500 was found.

Taylor et al.(1982) obtained a semipurified enzyme from the P. saccharophila ATCC 15946 strain, about 10% of the protein being sucrose phosphorylase. Purification steps of an ultrasonically obtained cell extract included ammonium sulfate precipitation, Sephadex G-25 elution, and DEAE cellulose and Sephadex G-100 chromatography. This enzyme preparation was subsequently used in immobilization procedures. Specific sucrose phosphorylase activities and percentage recovery of purification methods are given in Table IV.

TABLE III

PURIFICATION OF P. saccharophila SUCROSE PHOSPHORYLASE ACTIVITY

Purification step	Specific activity[a] (units/mg protein)	Total activity (units)
Crude cell extract[b]	0.19	5200
Protamine sulfate precipitate	0.19	4400
Ammonium sulfate precipitation	0.32	3940
DEAE cellulose column	3.12	1850
Ammonium sulfate and dialysis	3.15	1660
Calcium phosphate gel	33.1	1630
Carboxymethyl cellulose column	39.6	1040
Sephadex column	59.1	400

[a]One unit = amount of enzyme which esterifies 1 μmol P_i per minute per milliliter under standard conditions of reaction.
[b]Obtained after ultrasonication of 1.1 kg cell paste and centrifuging cell walls.

C. EFFECT OF TEMPERATURE ON SUCROSE PHOSPHORYLASE ACTIVITY

Doudoroff (1943) demonstrated that the sucrose phosphorylase activity of P. saccharophila cell extracts at 30°C exactly doubled compared with that at 20°C, indicating a Q_{10} of 2 for the phosphorolysis reaction. At 40°C, 80% of the activity at 30°C was found, indicating partial denaturation. Exposure of cell extract for 10 minutes to 55°C reduced the original activity by 55%.

Irreversible denaturation of P. saccharophila sucrose phosphorylase occurs at temperatures above 40°C. About 23% of the L. mesenteroides sucrose phosphorylase was inactivated after 5 minutes at 47.5°C and 75% after 10 minutes; at 50°C, 70% of the activity was lost within 4 minutes and 88% within 8 minutes; complete denaturation occurred at 60°C after 1 minute (Weimberg and Doudoroff, 1954). Kelly and Butler (1980) checked the temperature stability of the L. mesenteroides NRRL B-1065 enzyme at 4 and 30°C; 12 and 32% losses were detected after 2 and 14 days, respectively. Purified frozen (−18°C) enzyme preparations kept their activity for several months (Weimberg and Doudoroff, 1954). The C. pasteurianum enzyme displayed optimal activity in a rather broad temperature interval between 30 and 40°C (Tsai et al. 1980). A typical temperature profile for crude L. mesenteroides ATCC

TABLE IV

SPECIFIC SUCROSE PHOSPHORYLASE ACTIVITY AND PERCENTAGE
RECOVERY OF PURIFICATION METHODS[a]

Specific activity[b] (units)		Recovery	
Leuconostoc mesenteroides	Pseudomonas saccharophila	Leuconostoc mesenteroides	Pseudomonas saccharophila
142	30.6	26%	55%
1320	—	10.2%	—
—	1180 (homogeneous enzyme)	—	7.7%
980	—	42%	—

[a]References (in order of results shown in table): Weimberg and Doudoroff (1954); Doudoroff (1955); Silverstein et al. (1967); and Chassy and Krichevsky (1972).
[b]The specific activity is expressed in units as defined by Doudoroff (1943): one unit = the amount of enzyme which esterifies 1 μmol P_i per 20 minutes in a solution of 0.1 M sucrose and 0.033 M P_i at 30°C and pH 6.64.

12291 sucrose phosphorylase activity (P_2O_5-dried cells as enzyme source) is given in Fig. 5. Temperature optima of different sucrose phosphorylases are given in Table V. It can generally be concluded that once the critical temperature of about 40°C is surpassed, rapid denaturation of the enzyme occurs; this is a limiting factor in considering industrial applications of this enzyme.

D. EFFECT OF pH ON THE SUCROSE PHOSPHORYLASE REACTION

As early as 1943, Doudoroff demonstrated that pH influenced the equilibrium of the sucrose phosphorylase reaction: at pH 6.6, an equilibrium constant (K_{eq}) of 0.05 was found; at pH 5.8, the K_{eq} value was 0.09. The pH effect on enzyme activity has been widely tested with a range of buffer systems, as summarized in Table VI. Phosphate buffer systems are not ideal, since phosphate is a substrate for the enzyme. Grazi et al. (1977) found a sharp pH optimum at 7.5 for the P. saccharophila enzyme, whereas Tsai et al. (1980) found a broad pH optimum between 6.3 and 7.5, with pH 4.5 and 9.5 as limits for the C. pasteurianum enzyme. A typical pH activity profile of crude L. mesenteroides ATCC 12291 extract and P_2O_5-dried cells is given in Fig. 6. A survey of physicochemical properties of sucrose phosphorylase is given in Table V.

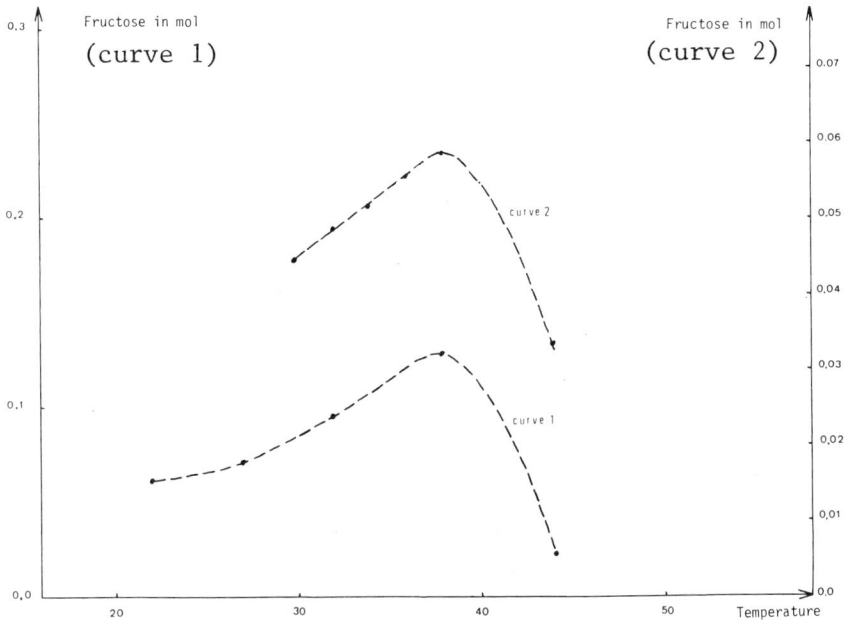

FIG. 5. Influence of temperature on sucrose phosphorylase activity. Curve 1: P_2O_5-dried *Leuconostoc* cells as enzyme source. Curve 2: Extract as enzyme source (De Laporte, 1982).

E. Substrate Specificity of Sucrose Phosphorylase

Sucrose phosphorylase catalyzes the transfer of glucose from sucrose or glucose-l-phosphate to an acceptor (inorganic phosphate or fructose, respectively); the enzyme is considered as a transglycosidase.

Sucrose phosphorylase displays a high substrate specificity toward the glycosyl donors; with inorganic phosphate or fructose as acceptor, only sucrose and glucose-1-phosphate were able to transfer their glucose moiety (Table VII). However, Gold and Osber (1971) found that in the phosphorolysis reaction, sucrose can be replaced by glucose-1-fluoride, with formation of glucose-1-phosphate and fluoride. The reverse reaction was not observed. As to the glycosyl acceptor, substrate specificity is rather low, as seen from the data compiled in Table VIII. Glucose can be transferred to a wide range of acceptors, which can be divided into two groups: (1) good acceptors, i.e., H_3PO_4, H_3AsO_4, and sugars, and (2) poor acceptors, i.e., H_2O and alcohols. The good acceptors can further be subdivided into inorganic compounds (H_3PO_4 and H_3AsO_4), keto sugars (D-fructose, L-sorbose, D-xylulose, L-ribulose, and D-rhamnulose), and the aldo sugar L-arabinose. Arsenate

TABLE V

PHYSICOCHEMICAL PROPERTIES OF SUCROSE PHOSPHORYLASES

Producing organism	Molecular weight	Specific volume	Sedimentation constant (S)	Optimal temperature (°C)	Optimal pH
Pseudomonas saccharophila	78,000[a]	0.74[b]	—	—	7.5[e]
	84,000[b]	—	5.2[b]	40 c,d	6.6–6.8[f]
	50,000[g]	—	—	—	6.4–7.0[d]
Pseudomonas putrefaciens	—	—	—	—	6.6–6.8[f]
Leuconostoc mesenteroides	—	—	—	35[g]	6.6–6.8[f]
				38[j]	6.35–6.52[i]
Clostridium pasteurianum	36,500[a]	—	—	30–40[a]	6.3–7.5[a]
Pullularia pullulans[h]	—	—	—	—	—

[a]Tsai et al. (1980).
[b]Silverstein et al. (1967).
[c]Doudoroff (1955).
[d]Doudoroff (1943).
[e]Grazi et al. (1977).
[f]Weimberg and Doudoroff (1954).
[g]Mieyal and Abeles (1970).
[h]Imshenetskii and Kondrat'eva (1954).
[i]Kelly and Butler (1980).
[j]De Laporte (1982).

TABLE VI

SURVEY OF BUFFER SYSTEMS USED FOR SUCROSE PHOSPHORYLASE ACTIVITY MEASUREMENTS

Buffer	pH	Microorganism	Reference
0.033 M phosphate–0.1 MNaCO$_3$	6.64[a]	P. saccharophila	Doudoroff, (1943)
	6.4–7.0[b]	P. saccharophila	Doudoroff (1943)
0.133 M acetate	6.85[a]	P. saccharophila	Hassid et al. (1944)
0.33 M Sörensen phosphate	6.64[a]	L. mesenteroides	Doudoroff (1955)
		P. pullulans	Imshenetskii and Kondrateva (1980)
	6.6–6.8[b]	L. mesenteroides	
		P. saccharophila	Weimberg and Doudoroff (1954)
		P. putrefaciens	
0.033 M citrate	6.64[a]	P. saccharophila	Abraham and Hassid (1957)
0.033 M tris-maleate	7.0[a]	P. saccharophila	Silverstein et al. (1967)
	6.3–7.5[b]	C. pasteurianum	Tsai et al. (1980)
0.1 M tris-maleate	6.35[a]	P. saccharophila	Mieyal et al. (1972)
0.05 M tris-HCl	7.5[a]	L. mesenteroides	Chassy and Krichevsky (1972)
0.1 M phosphate–0.05 M tris–0.05 M acetate	7.5[b]	P. saccharophila	Grazi et al. (1977)
0.02 M imidazol	6.34–6.52[b]	L. mesenteroides	Kelly and Butler (1980)
0.025 M MES buffer	6.5[a]	P. saccharophila	Mieyal et al. (1972)
0.040 M MES	6.9	P. saccharophila	Taylor et al. (1982)
0.1 M MOPS	6.8	L. mesenteroides	De Laporte et al. (1982)
0.05 M tris-maleate	6.2	L. mesenteroides	Daurat-Larroque et al. (1982)

[a] pH used in experiments.
[b] pH optimum.

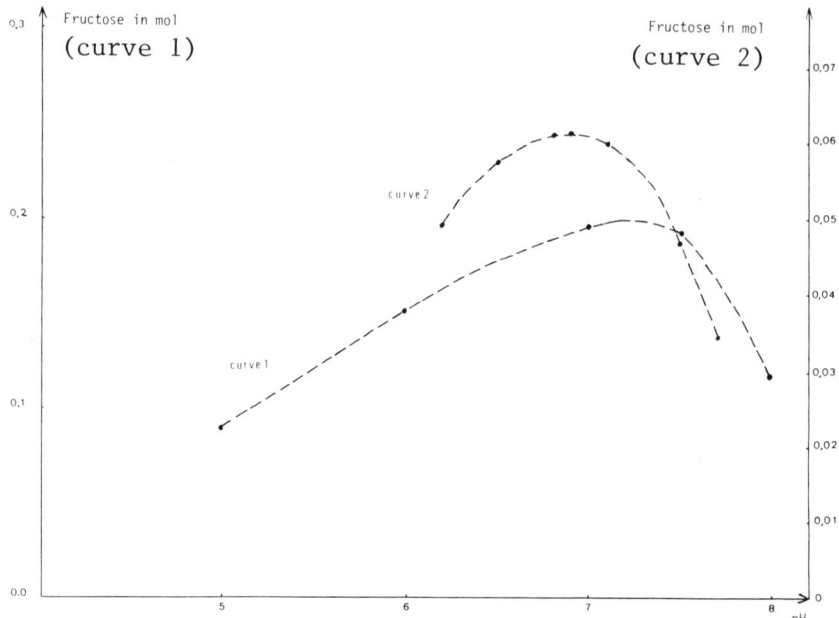

FIG. 6. Effect of pH on *Leuconostoc mesenteroides* sucrose phosphorylase. Curve 1: P_2O_5-dried cells as enzyme source. Curve 2: Enzyme extract as enzyme source (De Laporte, 1982).

can easily substitute for phosphate in biochemical reactions, though the esters produced are unstable compared to phosphate esters.

The disaccharides formed from the good acceptor keto sugars all correspond to the following structural formula:

Keto sugar	R^1	R^2	R^3	R^4
D-Fructose	H	CH_2OH	OH	H
L-Sorbose	CH_2OH	H	OH	H
D-Xylulose	H	H	OH	H
L-Ribulose	H	H	H	OH
D-Rhamnulose	H	CH_3	OH	H

The 3-α-D-glucanopyranosyl-L-arabinose does not correspond to this general formula:

3- α - D - glucopyranosyl - L - arabinose

A certain similarity can be observed, however—the glucosyl acceptor always has a hydroxyl group in a cis position relative to the glucosidic bond. Compounds unable to serve as glycosyl acceptor are D-glucose, D-ribose, D-xylose, D-mannose, D-lyxose, D-galactose, D-arabinose, D-tagatose, D-ribulose, D-fructose-6-phosphate, D-fructose-1,6-diphosphate, D-mannoheptulose, L-fructose, L-rhamnose, L-xylulose, turanose, dihydroxyacetone, erythrulose, and L-fucose (Doudoroff et al., 1947b). It is strange that D-tagatose, which corresponds to the general keto sugar structure, is not able to react with α-D-glucose-1-phosphate. The occurrence of D-tagatose in the pyranose form and not in the furanose form could explain this anomaly. Furthermore, it is striking that, based on differences in glucose acceptor specificity, the sucrose phosphorylases from *P. saccharophila* and *L. mesenteroides* can be discerned from that from *P. putrefaciens* (Weimberg and Doudoroff, 1954) (Table IX).

F. Inhibition and Inactivation of Sucrose Phosphorylase

Several compounds inhibit sucrose phosphorylase activity. Glucose displays a strong inhibitory action on sucrose phosphorolysis as well as on its synthesis (Fig. 7). Activity loss can be compensated by increasing the sucrose concentration, but phosphate has no effect. Doudoroff (1943) concluded from these observations that glucose and sucrose compete for the same active site on the enzyme (competitive inhibition). The presence of raffinose, trehalose, maltose, mannose, galactose, and NaF hardly influenced sucrose phosphorylase. Phlorizin [1,2 -β-D-glucopyranosyloxy)-4,6-(dihydroxyphenyl)-3-(4-hydroxyphenyl)-1-propanon] had an inhibitory effect only on sucrose synthesis and not on phosphorolysis. Xylose, xylose-1-phosphate, arabinose, galactose, and

TABLE VII

SPECIFICITY OF SUCROSE PHOSPHORYLASE FOR GLUCOSYL DONORS

Glycosyl donor	Acceptor	Formed product
Phosphorolysis		
Sucrose	$P_i{}^a$	Glucose-1-phosphate + fructose
Glucose-1-fluoride	$P_i{}^b$	Glucose-1-phosphate + F$^-$
Raffinose (melibiose-1-fructoside)	$P_i{}^a$	—
Trehalose	$P_i{}^a$	—
Maltose	$P_i{}^a$	—
Glycogen	$P_i{}^a$	—
Starch	$P_i{}^a$	—
Melezitose	$P_i{}^c$	—
Synthesis		
α-D-Glucose-1-phosphate	Fructose^a	Sucrose + P_i
α-D-Melibiose-1-phosphate	Fructose^d	—
α-D-Xylose-1-phosphate	Furctose^e	—
α-D-Maltose-1-phosphate	Fructose^e	—
α-D-Galactose-1-phosphate	Fructose^e	—
α-D-Mannose-1-phosphate	Fructose^e	—
α-L-Glucose-1-phosphate	Fructose^e	—
α-D-Glucose-1-phosphate	NaF^b	—

[a]Doudoroff et al. (1943).
[b]Gold and Osber (1971).
[c]Weimberg and Doudoroff (1954).
[d]Doudoroff et al. (1947b).
[e]Hassid and Doudoroff (1950).

galactose-1-phosphate only weakly inhibit sucrose phosphorylase as compared to glucose.

Silverstein et al. (1967) found that the glycosyl acceptors (phosphate, arsenate, and fructose) might cause inhibition. Ethylene glycol noncompetitively inhibits phosphorolysis; trancyclohexanediol is a competitive inhibitor toward phosphate and is a noncompetitive one toward sucrose (Mieyal et al., 1972). Table X presents a summary of sucrose phosphorylase inhibitors with their dissociation constants K_i. Solvents other than water, such as acetonitrile and dioxane, and compounds such as chloroethanol and mercaptoethanol irreversibly denature the enzyme (Mieyal et al., 1972), as do carbodiimide and glycine ethyl ester (DeToma and Abeles, 1970) and NaIO$_4$ (10^{-4} M) (Voet and Abeles, 1970).

TABLE VIII

SPECIFICITY OF SUCROSE PHOSPHORYLASE FOR GLUCOSYL ACCEPTORS

Glucosyl donor	Glucosyl acceptor	Formed products
	Synthesis	
α-D-Glucose-1-P	Good acceptors	
	D-Fructose[a]	α-D-glucopyranosyl-(1→2)-β-D-fructofuranoside + P_i (= sucrose)
	L-Sorbose[b]	α-D-glucopyranosyl-(1→2)-α-L-sorbofuranoside + P_i
	D-Xylulose[c]	α-D-glucopyranosyl-(1→2)-β-D-xyloketofuranoside + P_i
	L-Ribulose[c]	α-D-glucopyranosyl-(1→2)-α-L-arabinoketofuranoside + P_i
	D-Rhamnulose[e]	α-D-glucopyranosyl-(1→2)-β-D-rhamnoketofuranoside + P_i
	L-Arabinose[c]	α-D-glucopyranosyl-(1→3)-L-arabinofuranoside + P_i
	Phosphorolysis	
Sucrose	Good acceptors	
	H_3PO_4[a]	Glucose-1-phospate + fructose
	H_3AsO_4[d]	Glucose-1-arsenate + fructose
	Poor acceptors	
	H_2O[f]	Glucose + fructose
	Methanol[g]	α-Methylglucoside + fructose
	Ethanol[g]	Ethylglucoside + fructose
	cis-1,2-Cyclohexanediol[g]	Hydroxycyclohexylglucoside + fructose
	trans-1,2-cyclohexanediol[g]	Hydroxycyclohexylglucoside + fructose
	Ethyleneglycol[g]	Hydroxyethylglucoside + fructose

[a] Doudoroff et al. (1943.)
[b] Doudoroff et al. (1944).
[c] Doudoroff et al. (1947b).
[d] Doudoroff et al. (1947c).
[e] Palleroni and Doudoroff (1956).
[f] Weimberg and Doudoroff (1954).
[g] Mieyal et al. (1972).

TABLE IX

DIFFERENCES IN SUCROSE PHOSPHORYLASE GLUCOSE ACCEPTOR SPECIFICITY

		Glucosyl acceptor specificity	
Glucosyl donor	Glucosyl acceptor	L. mesenteroides and P. saccharophila	P. putrefaciens
α-D-Glucose-1-P	L-Sorbose	+	−
α-D-Glucose-1-P	D-Xylulose	+	−

V. Kinetics of the Sucrose Phosphorylase Reaction

Early experiments with labeled inorganic phosphate added to glucose-1-phosphate and sucrose phosphorylase, in the absence of fructose, revealed a rapid exchange of phosphate, which led Doudoroff et al. (1947a) to propose the following reaction scheme:

$$\text{glucose-1-phosphate} \rightleftharpoons \text{glucose–enzyme} + \text{phosphate}$$
$$\underline{\text{glucose–enzyme} + \text{phosphate}^* \rightleftharpoons \text{glucose-1-phosphate}^*}$$
$$\Sigma \; \text{glucose-1-phosphate} + \text{phosphate}^* \rightleftharpoons \text{glucose-1-phosphate} + \text{phosphate}^*$$

The formation of a glucose–enzyme complex was substantiated by the fact that sucrose, in the presence of L-sorbose and the enzyme without inorganic phosphate, was converted into glucose-L-sorboside and fructose:

$$\text{sucrose} + \text{enzyme} \rightleftharpoons \text{glucose–enzyme} + \text{fructose}$$
$$\underline{\text{glucose–enzyme} + \text{L-sorbose} \rightleftharpoons \text{glucose-L-sorboside} + \text{enzyme}}$$
$$\Sigma \; \text{sucrose} + \text{L-sorbose} \rightleftharpoons \text{glucose-L-sorboside} + \text{fructose}$$

These observations led to the following reaction mechanism:

$$\text{sucrose} + \text{enzyme} \xrightleftharpoons{+ \text{fructose}} \text{glucose–enzyme}$$
$$\updownarrow + \text{phosphate}$$
$$\text{glucose-1-phosphate} + \text{enzyme}$$

Silverstein et al. (1967) isolated the glucose–enzyme complex in denaturated form. Sucrose labeled in the glucose moiety was added to purified sucrose phosphorylase and the mixture was denaturated with boiling methanol or by bringing the pH to 3.0. Glucose (1.5–1.9 mol per mole of enzyme) was formed, which indicates the existence of two

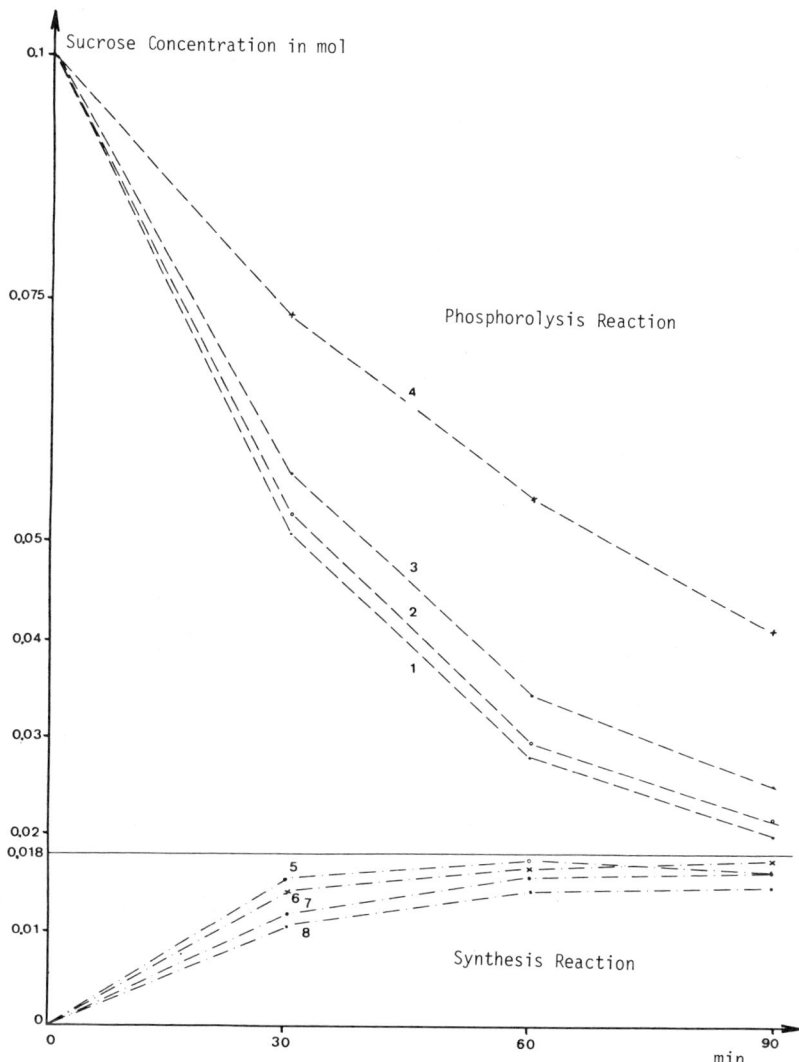

FIG. 7. Inhibitory effect of glucose on sucrose phosphorylase activity. Curves 1 and 5, no glucose added; 2 and 6, 0.001 mol glucose added; 3 and 7, 0.01 mol glucose added; and 4 and 8, 0.1 mol glucose added. Curves 1–4 represent phosphorolysis and curves 5–7 represent synthesis (De Laporte, 1982).

TABLE X

SURVEY OF DIFFERENT SUCROSE PHOSPHORYLASE INHIBITORS
AND THEIR DISSOCIATION CONSTANTS

Donor (G–X)	Acceptor (Y)	Inhibitor (I)	$K^i \times 10^3$ (M)
Sucrose	P_i	Glucose	0.46^a
Sucrose	P_i	1,5-Gluconolactone	0.62^b
Sucrose	P_i	P_i	250^a
Sucrose	P_i	Ethylene glycol	2500^c
Sucrose	P_i	trans-1,2-Cyclohexanediol	1400^c
			250^c
			400^c
Sucrose	P_i	cis-1,2-Cyclohexanediol	600^c
Sucrose	L-Sorbose	L-Sorbose	670^a
α-D-Glucose-1-phosphate	D-Fructose	d-Fructose	250^a
α-D-Glucose-1-phosphate	L-Sorbose	L-Sorbose	170^a

^{a}Silverstein et al. (1967).
^{b}Mieyal and Abeles (1970).
^{c}Meiyal et al. (1972).

active sites per mole of enzyme (Mieyal and Abeles, 1970). This fact, and molecular weight determinations with and without sodium dodecyl sulfate, suggest that the enzyme contains two subunits, each with an active site. Proteolysis of the denatured glucose–enzyme complex yielded peptides with covalently bound glucose in the β configuration. A survey of kinetic constants of the sucrose phosphorylase reaction is given in Table XI and a scheme of the mode of action of the enzyme is presented in Fig. 8.

The equilibrium constant favors the phosphorolysis of sucrose over its synthesis. Preliminary data of Kelly and Butler (1980) suggest that ionic barium altered the equilibrium in the opposite direction. A detailed initial rate kinetic study with purified sucrose phosphorylase from P. saccharophila has indicated that the reaction proceeds via the ping-pong mechanism (Silverstein et al., 1967; Taylor et al., 1982). This has also been substantiated by isotopic exchange experiments (Doudoroff et al., 1947a) and isolation of a β-linked glucosyl enzyme intermediate, indicating that two inversions of glucose configuration occur during the course of the reaction (Voet and Abeles, 1970). Slight deviations from this classical reaction mechanism have been observed independently by De Laporte (1982) and by Taylor et al. (1982) during phosphorolysis, especially at high phosphate concentrations. De Laporte (1982) studied and reported results on the kinetics of the L. mesenteroides ATCC 12291 enzyme (De Laporte, 1982).

TABLE XI

SURVEY OF KINETIC CONSTANTS OF THE SUCROSE PHOSPHORYLASE REACTION

Donor (G–X)	Acceptor (Y)	$V_{max} \times 10^3$ (mmol/min/unit)	$k_{G-X} \times 10^e$ (M)	$K_Y \times 10^3$ (M)	$K_i \times 10^3$ (M)	Ref.[a]
Sucrose	P_i	1.9	25.4	21.3	137	(6)
Sucrose	P_i	1.4	2.5	1.8	250	(1)
Sucrose	P_i	1.28	0.92	—	—	(2)
Sucrose	P_i	—	1.66	0.33	—	(3a)
Sucrose	P_i	—	3.3	2	—	(3b)
Sucrose	AsO_4^{3-}	1.6	2.6	1.2	—	(3b)
Sucrose	L-Sorbose	0.78	1.6	110	67	(1)
Sucrose	H_2O	0.025	0.50	—	—	(1)
Sucrose	Ethyleneglycol	0.053	—	870	—	(1)
Sucrose	Ethyleneglocol	0.046	—	2700	—	(4)
Sucrose	trans-1,2-cyclohexanediol	0.023	—	400	—	(4)
Sucrose	trans-1,2-cyclohexanediol	0.014	—	270	—	(4)
Sucrose	Methanol	0.038	—	2300	—	(4)
Sucrose	Methanol	0.011	—	2100	—	(4)
Sucrose	cis-1,2,-Cyclohexanediol	0.0046	—	390	—	(4)
Sucrose	Ethanol	0.023	—	20,000	—	(4)
Sucrose	H_2O	0.005	—	20,000	—	(4)
α-D-Glucose-1-P	AsO_4^{3-}	2.1	4.2	2.3	—	(1)
α-D-Glucose-1-P	D-Fructose	0.91	2.3	13	250	(1)
α-D-Glucose-1-P	L-Sorbose	0.62	3.2	130	170	(1)
α-D-Glucose-1-P	H_2O	0.021	0.46	—	—	(1)
α-D-Glucose-1-P	P_i	3.4	1.17	—	—	(2)
α-D-Glucose-1-P	H_2O	0.023	1	—	—	(2)

[a] (1), Silverstein et al. (1967); (2), Gold and Osbern (1971); (3), Grazi et al. (1977); a, free enzyme and b, immobilized enzyme; (4), Mieyal et al. (1972); (5), Taylor et al. (1982); (6), De Laporte (1982).

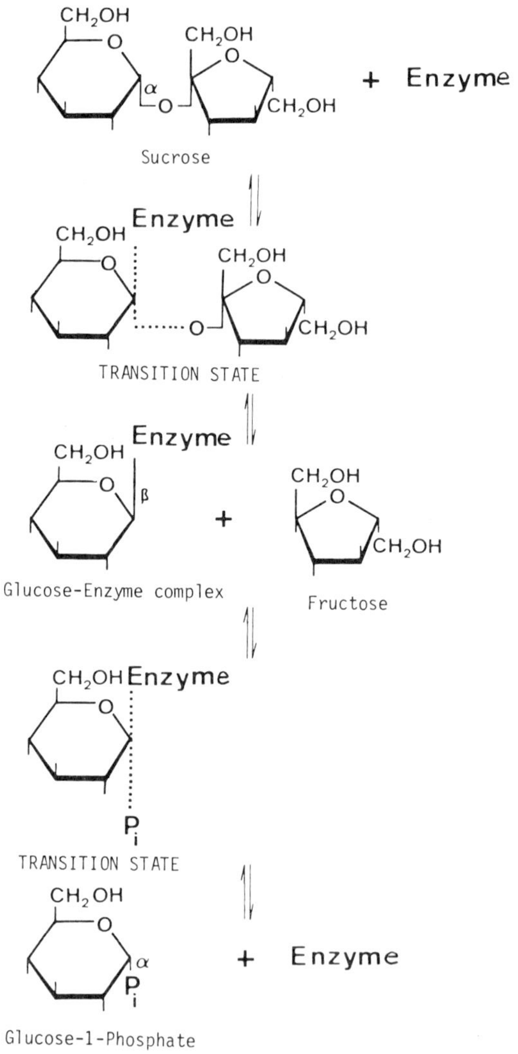

FIG. 8. Mode of action of sucrose phosphorylase.

VI. Immobilization of Sucrose Phosphorylase

Enzyme and cell immobilization have received much attention in view of industrial applications of biocatalysts (Vandamme, 1976, 1981, 1983; Abbott, 1976; Chibata and Tosa, 1981; Brodelius and Vandamme, 1987). Grazi et al. (1977) were the first to immobilize a semipurified preparation of the P. saccharophila sucrose phosphorylase. The enzyme was entrapped in cellulose triacetate fibers following the procedure of Dinelli (1972).

About 40% of the original enzyme activity was retained, the drop in activity mainly being caused by diffusional limitation of the substrate. The immobilized biocatalyst retained all activity for several months at 20°C. At 37°C at pH 6.0, activity remained constant for several days. It appeared that the cellulose matrix confers excellent protection against denaturation or microbial contamination. The immobilized enzyme displayed a broad pH optimum between pH 6.0 and 8.0; the free enzyme displayed peak optimal activity at pH 7.5.

Taylor et al. (1982) reported on the kinetics of immobilized sucrose phosphorylase of P. saccharophila. A semipurified enzyme preparation was immobilized on porous ceramic beads with 3-aminopropyl-triethoxysilane and glutaraldehyde; one-third of the original enzyme activity remained after immobilization; pore diffusional resistance was negligible under laboratory conditions. The enzyme half-life was 35 days at 30°C, but only 5 days at 10°C. The pH optimum was between 6.5 and 7.0. An activation energy of about 12.5 kcal/mol was calculated. Kinetic constants were in good agreement with those for the soluble enzyme.

The L. mesenteroides ATCC 12291 sucrose phosphorylase has also been immobilized (De Laporte, 1982; Daurat-Larroque et al., 1982; Van Loo, 1983; Machtelinckx, 1983; Vandamme et al., 1984a,b). Several immobilization procedures were tested so as to arrive at a practical immobilized sucrose phosphorylase biocatalyst. Inclusion techniques using P_2O_5-dried cells or enzyme in alginate, carrageenan, and gelatin (all with or without cross-link to glutaraldehyde), in alginate/gelatin or gelatin/tannin mixtures were tested by Van Loo (1983). Leuconostoc mesenteroides cells, harvested from the fermentation broth at their peak sucrose phosphorylase level, were P_2O_5-dried and immobilized in the form of beads. Only entrapment in gelatin, combined with a minimal glutaraldehyde cross-linking, yielded a successful biocatalyst. Residual enzyme activity after this inclusion–immobilization procedure averaged about 45%; this activity drop is mainly a result of protein cross-linking with glutaraldehyde. The pH optimum of this catalyst was comparable to that of the free enzyme (6.8 to 7.6), but full activity was seen over a broader pH range (4 to 9) (Fig. 9) (Van Loo, 1983). The temperature effect on enzyme activity revealed an optimum of 44°C after 10 minutes of reactions. This temperature–activity optimum decreased with increasing reaction periods (35°C after 30 minutes and 27°C after 2 hours).

Optimization of the gelatin-immobilization procedure and of the storage of the "enzyme" beads led to the successful use of such an

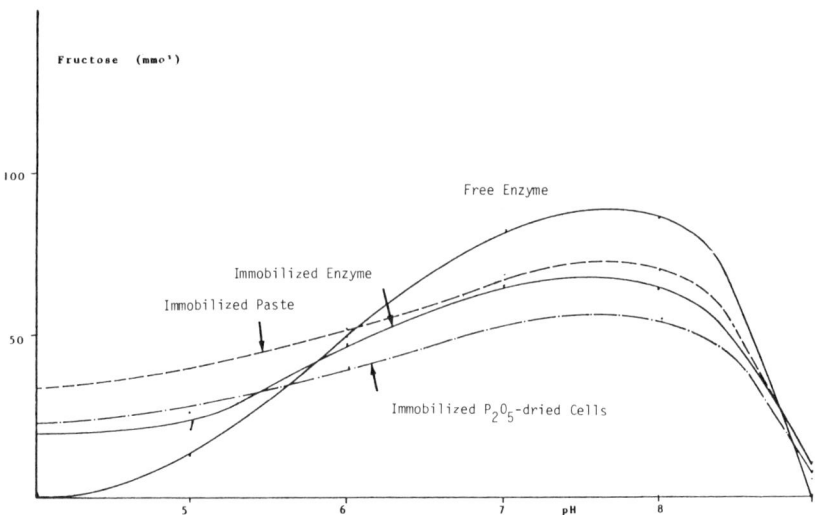

FIG. 9. Effect of pH on activity of free and immobilized sucrose phosphorylase (Van Loo, 1983).

immobilized biocatalyst in continuous-column reactors to convert sucrose and inorganic phosphate to glucose-1-phosphate and fructose (Machtelinckx, 1983; Van Loo, 1983). Phosphorolysis of sucrose usually approached about 60%, although maximum conversion (80%) could be obtained. At the start of the continuous reaction, specific productivity averaged 15 to 20 mmol fructose/hour/g active P_2O_5-dried cells. This productivity dropped considerably upon further operation; leakage of cells from the beads and thermal denaturation of the sucrose phosphorylase ($t_{1/2}$ = 23 to 87 hours) are responsible for this negative aspect (Fig. 10). Reaction time–temperature combinations revealed an inverse relationship between optimal temperature and prolonged continuous reaction. Increasing the flow rate resulted initially in an increase in productivity, but lower flow rates proved better upon prolonged reaction. High substrate concentrations yielded a relatively low degree of conversion, but the reaction rate was higher.

Column geometry contributes appreciably to the efficiency of the sucrose conversion reaction (Machtelinckx, 1983). Short half-life values as well as considerable cell leakage limit industrial use of gelatin-immobilized sucrose-phosphorylase-rich cells; furthermore, the compressibility of gelatin beads prevents scaling-up in large-column bioreactors. These handicaps could be partially overcome by increased cross-linking, but this reduces enzyme activity considerably.

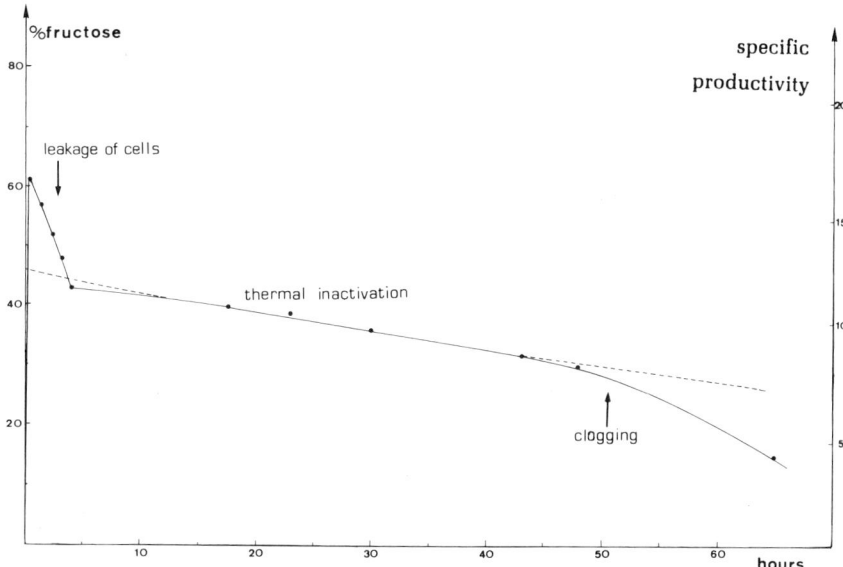

FIG. 10. Operational stability of immobilized sucrose phosphorylase column bioreactor. Reaction conditions for sucrose phosphorolysis: temperature, 37 °C; column H/D ratio, 8.7; flow rate, 75 ml/h (dilution rate = 4.2/h; substrate concentration, 0.1 M sucrose + 0.1 M phosphate (pH 7.3). Initial degree of conversion, 60%; initial velocity (specific productivity), 17 mmol fructose per hour per gram active P_2O_5 cells; half-time value, 57 h (experimental) and 87 h (theoretical, dashed line); loss of column performance due to leakage of cells, bead compression in column, and thermal inactivation (Machtelinckx, 1983).

An interesting cell immobilization procedure was proposed by researchers at the Laboratoria Recherche Processi Microbiologia, Monterotendo, Italy (S. Giovenco; personal communication). Fermentation broth is treated with polyethyleneimine and glutaraldehyde; this results in the *in situ* formation of stable *Leuconostoc* cell aggregates, which are subsequently entrapped in cellulose acetate beads. Also, Larroque et al. (1981) and Daurat-Larroque et al. (1982) studied sucrose phosphorylase immobilization in the hope of increasing sucrose synthesis capacity. The enzyme from *L. mesenteroides* ATCC 12291 was coupled through glutaraldehyde to porous glass and fitted into a column. A second column, containing activated coal (which displays higher affinity for sucrose than for fructose and glucose-1-phosphate (Whistler and Durso, 1950), was connected to the outlet of the immobilized enzyme column. The eluate from this column was recirculated through the first column; in this way, the enzyme was continuously confronted with high substrate concentration and low product concentration, while formed sucrose was constantly adsorbed onto the charcoal column. This system might represent an example for the industrial performance of enzyme reactions in an energetically unfavorable direction. A partially purified enzyme (sucrose phosphorylase) from L. mesenteroides contained in a hollow-fiber device was able to produce low levels (1%) of sucrose from fructose and glucose-1-phosphate at pH 6.2 over a 4-day period (Daurat-Larroque et al., 1982). This sucrose phosphorylase was coimmobilized with starch phosphorylase (EC 2.4.1.1) (which converts starch and P_i to glucose-1-phosphate) on DE-52 cellulose columns. Since the pH optima of both phosphorylases differ considerably, a mean pH value of 6.2 was chosen to operate the column, with 2% starch, 20 mM K_2HPO_4, and 300 mM fructose in 20 mM maleate buffer; an average of 2 mM sucrose was obtained in the eluate at a flow rate of 5.4–11.0 ml/hr. Since high-ionic-strength media (high phosphate) desorbed the enzymes from the DE-52 cellulose columns, coimmobilization based on hydrophobic interactions was evaluated with the enzymes covalently attached to Sephadex 4B. The purified L. mesenteroides sucrose phosphorylase was also covalently bound to aminopropyl silanylated glass beads, activated by glutardialdehyde, and was used for continuous sucrose synthesis from excess (300 mM) fructose and varying glucose-1-phosphate concentrations; an optimal yield of 14 mM sucrose at a beginning concentration of 26 mM glucose-1-phosphate was obtained, representing 54% conversion of this substrate (Daurat-Larroque et al., 1982). It is obvious that further research is needed in order to arrive at an ideal combination of good enzyme activity and high enzyme stability in continuous-column bioreactors.

VII. Applications of Sucrose Phosphorylase

A. Production of Sucrose from Starch

Since time immemorial, man has directed attention to sweeteners. Cane sugar has been long used, although beet sugar was not introduced until the nineteenth century. Today, world sucrose production reaches about 91 million tons, with an average of 40 kg per capita in the West, and an average of 20 kg worldwide. An increase in sucrose consumption is expected especially in developing countries. In India and China, with about 1.9 billion people and an average annual consumption of 6.4 and 4.1 kg of sucrose per capita, respectively, a steady increase in sucrose utilization is being observed (Danic-Careil,1979). Sucrose is and has always been a very controversial and politically loaded product. Small variations in production influence drastically the demand on a world scale and enormous price fluctuations have been experienced in the last decades. Due in part to this unstable market situation, attention has been diverted to alternative substrates for the production of sweeteners; rather automatically, starch has received and still gets full attention in this respect, e.g., production of glucose/fructose syrups, in addition to other developments (aspartame, thaumatin, and acesulfame).

However, the microbial production of sucrose from starch is an alternative that so far has received very little attention. In such a process, starch is enzymatically (amylase, EC 3.2.1.1; amyloglucosidase, EC 3.2.1.3) hydrolyzed to glucose, which is subsequently isomerized by glucose isomerase [D-(+)-xylose isomerase, EC 5.3.1.5] into fructose (Vandamme et al., 1981), which is the first substrate for sucrose phosphorylase. Pure fructose could also be derived from inulin enzymatically in one step by inulinase (EC 3.2.1.7) activity. Inulin is an abundant reserve material in several important agricultural crops (Fleming and Grootwassink, 1979; Vandamme and Derycke, 1983). The other substrate, glucose-1-phosphate, can also be obtained from starch and inorganic phosphate through activity of starch phosphorylase (EC 2.4.1.1.), an enzyme that is abundant in potatoes (Hollo et al., 1962). Both biotechnological reactions have been rather well studied, but independently from each other. The final step, the coupling of glucose-1-phosphate and fructose to sucrose and inorganic phosphate with sucrose phosphorylase has been studied very little (Butler et al., 1977); Larroque et al.,1982), yet it might become an important industrial bioreaction in certain countries. Inorganic phosphate is recycled, hence starch is the only input, with sucrose as the output (Fig. 11).

Sucrose has been produced at pH 6.8 at 30°C (13% yield) from 0.5% starch, 50 mM Na_2HPO_4 and 0.6 M fructose in a batch reactor by the combined action of starch phosphorylase and sucrose phosphorylase (Daurat-Larroque et al., 1982). Sucrose withdrawal to drive the reaction from the reaction medium (experimentally accomplished by adding invertase) increased the yield up to 25%, but this apparently resulted simultaneously in partial glucose inhibition of the sucrose phosphorylase (Weimberg and Doudoroff, 1954; Silverstein et al., 1967). Continuous sucrose formation with coimmobilized starch and sucrose phosphorylase has also been accomplished. Another sucrose-trapping system consists of a charcoal–celite column, connected to the outlet of a continuous immobilized sucrose phosphorylase reactor (Daurat-Larroque et al., 1982). Similarly, enzymatic hydrolysis of cellulose into cellobiose, with subsequent phosphorolysis to glucose-1-phosphate (and glucose) with cellobiose phosphorylase (EC 2.4.1.20), could provide a substrate for sucrose synthesis.

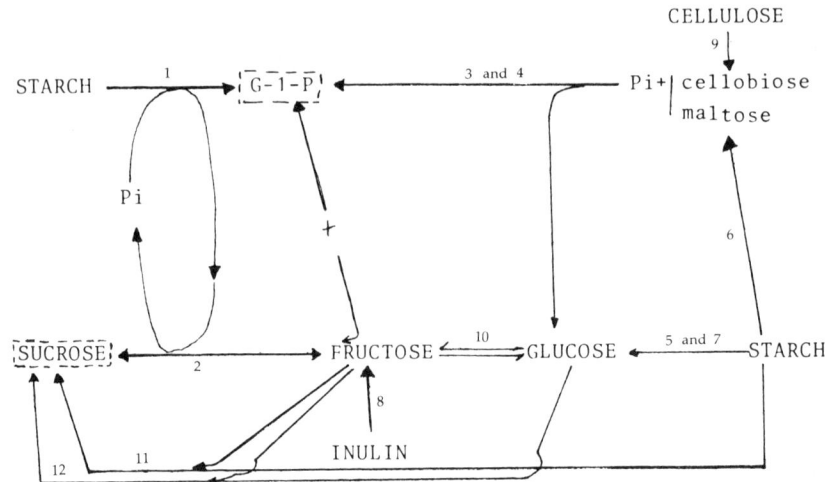

FIG. 11. Microbial enzymic production of sucrose. (1) Starch phosphorylase (EC 2.4.1.1); (2) sucrose phosphorylase (EC 2.4.1.7.); (3) cellobiose phosphorylase (EC 2.4.1.20); (4) maltose phosphorylase (EC 2.4.1.8); (5) amylase (EC 3.2.1.1); (6) β-amylase (EC 3.2.1.2); (7) amyloglucosidase (EC 3.2.1.3); (8) inulinase (EC 3.2.1.7); (9) cellulase (EC 3.2.1.4); (10) xylose isomerase (EC 5.3.1.5); (11) amylosucrase (EC 2.4.1.4); and (12) invertase (EC 3.2.1.26).

In general, renewable resources such as cellulose, starch, and inulin can be converted into sucrose via the above-mentioned biotechnological processes with the following important points to be stressed:

1. The sucrose phosphorylase is very specific in that the formation of another disaccharide from glucose-1-phosphate and fructose does not occur.
2. The sucrose phosphorylase reaction conserves the energy of the sucrose molecule, thus saving ATP or other energy sources.
3. Nonpathogenic L. mesenteroides species produce a stable enzyme with high activity under anaerobic (energy-saving) fermentation conditions at acid pH.
4. Immobilization procedures might improve the economics of this reaction.
5. Starch is abundant, notably in countries that have to import sucrose.
6. Sucrose is not only ubiquitous as a sweetening compound with functional properties, it is also used in unconventional sucro-chemical applications such as in production of polyurethane films and of sucrose esters used as emulsifying agents and surfactants (Parker, 1980).

In addition to sucrose phosphorylase action, the reverse reaction of a few other microbial enzymes that catalyze the in vivo breakdown of sucrose could be exploited for in vitro sucrose synthesis. (1) Reversed invertase action might be favored in a nonaqueous medium to couple glucose and fructose, but lack of substrate specificity of presently known invertases allows formation of nonsucrose fructosides (Butler et al., 1977). (2) Amylosucrase (EC 2.4.1.4), prepared from Neisseria perflava (also a producer of maltose phosphorylase, EC 2.4.1.8) (Selinger and Schramm, 1961), can form sucrose from starch and fructose in a single (reversed) enzymatic step (Hehre and Hamilton, 1946; Okada and Hehre, 1974; Daurat-Larroque et al., 1982); however, the very unstable nature of this enzyme ($t_{1/2}$ = 20 hours) and lack of good yields and safe microbial strains drastically limit its application at present.

B. Direct Production of Fructose and Glucose-1-Phosphate from Sucrose or Molasses

D-(−)-Fructose (or levulose) is a natural sugar. It is found in free form in most fruits and is the main sugar in honey; it occurs in many plants and fruits, bound with glucose, as sucrose. At present, its main application is in dietary foods. In pure crystalline form it is 1.8 times sweeter than sucrose, and is sweeter than all other naturally occurring sugars or sweet sugar alcohols. As a consequence, smaller quantities are needed to obtain an equal sweetening effect; thus it contributes fewer calories to the food and yet presents high sweetening efficiency. Fructose can also be used by diabetics and it is known not to contribute to dental decay (Koivistoinen and Hyvonen, 1980).

From a technological point of view, fructose displays attractive functional properties such as an improved expression of fruit aromas, which has led to numerous applications in manufacture of jams, canned foods, cold drinks, fruit juices, etc. In the baking process, fructose allows a desirable fast "browning," due to its susceptibility to the Maillard reaction, and its hygroscopicity contributes to the shelf life of bakery products. Furthermore, fructose displays, in combination with saccharin, a synergistic effect, and the bitter aftertaste of saccharin is masked.

It is generally assumed—in view of its above-mentioned properties—that the use of fructose will largely increase in the food industry in the next decades (Bus et al., 1977). At present, industrial fructose production occurs via fructose enrichment from invert sugar or from high-fructose corn syrups (HFCS). Invert sugar is obtained by enzymatic (invertase) or acid hydrolysis of sucrose to glucose and fructose. High-fructose corn syrups (42% fructose) are enzymatically produced from starch in a three-step process (α-amylase, amyloglucosidase, and D-(+)-xylose isomerase). From the glucose/fructose mixtures, fructose syrups (95% fructose) are prepared via chromatographic enrichment procedures. Such fructose syrups can also be obtained from plant fructose polymers such as inulin via acid or enzymatic (inulinase) hydrolysis (Fleming and Grootwassink, 1979; Vandamme and Derycke, 1983). Alternatively, fructose could be produced from sucrose with sucrose phosphorylase:

$$\text{sucrose} + P_i \rightleftharpoons \text{fructose} + \text{glucose-1-phosphate}$$

One obvious negative aspect is that this reaction is an equilibrium reaction. However, this process offers at least several important advantages:

1. Fructose and glucose-1-phosphate are easily separated, as compared to glucose/fructose mixtures.
2. The process can start with a pure material, sucrose; thus, recovery procedures are kept to a minimum
3. The fructose syrup crystallizes easily due to the absence of higher sugars, which inevitably are formed during the preparation of fructose from starch or inulin.
4. Glucose-1-phosphate is also produced and must be valorized as such or hydrolyzed to glucose and inorganic phosphate, which can then be recycled in the reaction.

C. Large-Scale Production and Use of Glucose-1-Phosphate

Production of glucose-1-phosphate is a scarely documented process, although this compound is a common component of all living matter. Direct isolation of glucose-1-phosphate from yeast has been tried, apparently with low yield (Miklailova and Bashkovich, 1969). Use of starch phosphorylase (EC 2.4.1.1) to convert starch to glucose-1-phosphate is another possibility that has received attention (Hollo et al., 1962; Bot and Dosa, 1967; Silonova and Lisovskaya, 1975). An E. coli enzyme has been described that transforms phosphoramidate and glucose into glucose-1-phosphate and the corresponding amine (Fujimoto and Smith, 1960). However, use of disaccharide (sucrose) phosphorylase as mentioned above seems the most promising means for glucose-1-phosphate production. Also, organic chemical syntheses are laborious compared to the enzymatic production methods. Potential applications of glucose-1-phosphate are mainly medically oriented (Ficini, 1956).

The enzyme and its derivatives display vasodilatory properties and can be added to blood-infusion solutions. Ca^{2+} (and amine) salts of glucose-1-phosphate are cardiotonic, could be used as psychostimulatory compounds and as tonicum, and stimulate Ca^{2+} deposition in the skeleton and teeth. In vinyl chloride polymerization processes, technical application may consist of coating the reactor walls with glucose-1-phosphate, to prevent polymer precipitation. Furthermore, glucose-1-phosphate can be used as starting material for glucuronic acid and glucaric acid (Ca-complexant) synthesis. Surfactant production based on glucose-1-phosphate is another interesting proposal (H. Vander Wiel, personal communication). Data on separation of mixtures of glucose-1-phosphate, sucrose, P_i, and fructose were reported by Hokse (1983).

D. Enzymatic Assay of Sucrose or Inorganic Phosphate

An enzymatic assay for sucrose or inorganic phosphate determination could be envisaged as follows:

$$\text{Sucrose} + P_i \xrightleftharpoons{\text{sucrose phosphorylase}} \text{glucose-1-phosphate} + \text{fructose}$$

$$\text{Glucose-1-phosphate} \xrightleftharpoons{\text{phosphoglucose isomerase}} \text{glucose-6-phosphate}$$

$$\text{Glucose-6-phosphate} + NADP^+ \xrightleftharpoons{\text{glucose-6-phosphate dehydrogenase}}$$
$$\text{6-phosphogluconic acid} + NADPH + H^+$$

Such a method would be particularly suitable for the assay of inorganic phosphate in the presence of labile phosphate esters; indeed, inorganic phosphate is transformed into glucose-1-phosphate in the presence of excess sucrose, under mild conditions, leaving labile phosphate esters intact. In the above reaction sequence, the formation of $NADPH + H^+$ (measured at 365 nm) is proportional to the inorganic phosphate level.

E. Synthesis of Labeled Sucrose

Sucrose phosphorylase can be used to synthesize sucrose, labeled in the glucose or in the fructose moiety:

$$*\text{Glucose-1-phosphate} + \text{fructose} \rightleftharpoons *\text{glucose-1-fructoside} + \text{phosphate}$$

or

$$\text{Glucose-1-phosphate} + *\text{fructose} \rightleftharpoons \text{glucose-1-}*\text{fructoside} + \text{phosphate}$$

These reactions using a ^{14}C label have been carried out with the *P. saccharophila* enzyme (Wolochow et al., 1949; Abraham and Hassid, 1957) as well as with *L. mesenteroides* sucrose phosphorylase (Chassy and Krichevsky, 1972).

These two types of labeled sucrose can contribute to the elucidation of the sucrose metabolic pathways in plants and microorganisms, fundamentally as well as practically, e.g., metabolism of sucrose by oral flora and its relation to tooth decay.

F. Production of Unnatural Sugars or Novel Disaccharides

As already mentioned, sucrose phosphorylase is not strictly specific as to glycosyl acceptors. As a result, the following novel disaccharides so far not encountered in nature have been synthesized (Table XII):

1. α-D-Glucopyranosyl-α-L-sorbofuranoside (Hassid et al., 1945, 1948; Mozza et al., 1975).
2. α-D-Glucopyranosyl-β-D-ketoxylofuranoside (Hassid et al., 1945).
3. α-D-Glucopyranosyl-(1→3)-L-arabinopyranoside (Doudoroff et al., 1947b).

TABLE XII

Reaction of Substrates with Glucose-1-Phosphate in the Presence of Sucrose Phosphorylase[a]

Class of compound	Reacting compound	Nonreacting compound
Ketotriose	—	Dihydroxyacetone
Ketotetrose	—	Erythrulose
Aldopentose	—	D-Xylose
Aldopentose	L-Arabinose	D-Arabinose
Ketopentose	D-Ketoxylose	L-Ketoxylose
Ketopentose	L-Ketoarabinose	D-Ketoarabinose
Methylpentose	—	L-Fructose
Methylpentose	D-Rhamnulose ⇌ D-rhamnose[b]	L-Rhamnose
Aldohexose	—	D-Glucose
Aldohexose	—	D-Mannose
Aldohexose	—	D-Galactose
Ketohexose	D-Fructose	L-Fructose
Ketohexose	L-Sorbose	—
Ketohexose	—	D-Tagatose
Ketohexose	—	D-Mannoheptulose
Fructose derivatives	—	Fructose-6-phosphate
Fructose derivatives	—	Fructose-1,6-diphosphate
Fructose derivatives	—	Turanose (3-glucosido-D-fructose)
Fructose derivatives	—	Degraded levan

[a]Doudoroff et al. (1947a).
[b]Palleroni and Doudoroff (1956).

4. α-D-Glucopyranosyl-α-L-ketoarabinofuranoside (Doudoroff et al., 1947b).

5. α-D-Glucopyranosyl-β-D-rhamnoketofuranoside (Palleroni and Doudoroff, 1956).

Many other disaccharides might be "constructed"; this could lead to synthetic sugars with new and important functional properties related to food value, sweetening power, taste and aroma, color and hue, fermentation, structure and consistency, crystallization, conservation, humidity control, water activity regulation, and freezing-point depression and regulation, among others.

G. Formation of High-Energy P-Bonds without the Need for ATP

Glucose-1-phosphate is formed by sucrose phosphorylase activity from sucrose and P_i without the need for ATP. This concept should gain wider interest in the synthesis of energy-rich P_i compounds with biotechnological potential and in economizing on ATP use of the living cell. Cloning of sucrose phosphorylase activity in industrially useful microbial strains seems important in this perspective.

Acknowledgment

This research was partially sponsored by Belgian industry and government —IWONL–IRSIA.

References

Abbott, B. J. (1976). *Adv. Appl. Microb.* **20**, 203–257.
Abraham, S., and Hassid, W. Z. (1957). In "Methods in Enzymology" (S. P. Colowick and N. O. Kaplan, eds.), Vol. 4, pp. 502–505. Academic Press, New York.
Ayers, W. A. (1959). *J. Biol. Chem.* **234**, 2819.
Balasubramaniam, K., and Kannangara, P. N. (1982). *J. Natl. Sci. Coun. Sri Lanka* **10**, 169–180.
Belocopitow, E. and Marechal, L. R. (1970). *Biochim. Biophys. Acta* **198**, 151–154.
Bot, G., and Dosa, I. (1967). *Biochim. Biophys. Acta* **2**, 341–347.
Brodelius, P., and Vandamme, E. J. (1987). *Biotechnology* **7a** (in press).
Bus, W. C., Reuvekamp, H., and Seinkels, J. (1977). *Glucose Inf.* **11**, 2–18.
Butler, L. G., Squires, R. G., and Kelly, S. J. (1977). *Sugar Azucar* **72**, 31–32, 67–68.
Chassy, B. M., and Krichevsky, M. I. (1972). *Anal. Biochem.* **49**, 232–239.
Chibata, I., and Tosa, T. (1981). *Annu. Rev. Bioeng.* **10**, 197–216.
Danic-Careil, D. (1979). *Entreprises Agric.* **114**, 10–14.
Daurat-Larroque, S., Hammar, L., and Whelan, W. J. (1982). *J. Appl. Biochem.* **4**, 133–152.
De Laporte, A. (1982). Doctoral thesis, University of Ghent.
De Laporte, A., De Mey, L., and Vandamme, E. J. (1981). *Abstr. Eur. Congr. Biotechnol.* 2nd April 5–10, Eastbourne, England p. 237.
De Laporte, A., De Valck, L., and Vandamme, E. J. (1982). *Antonie van Leeuwenhoek, J. Microbiol. Serol.* **48**, 516–519.

De Laporte, A., and Vandamme, E. J. (1983). *Natuurwet. Tijdschr.* **65**, 167–213.
De Moss, R. D., Bard, R.C., and Gunsalus, I.C. (1951). *J. Bacteriol.* **62**, 499–502.
De Toma, F., Abeles, R. H. (1970). *Fed. Proc., Fed. Am. Soc. Exp. Biol.* **29**, 461.
De Valck, L. (1981). M. Sc. thesis, University of Ghent.
Dinelli, D. (1972). *Process Biochem.* **7**, 9–12.
Doudoroff, M. (1943). *J. Biol. Chem.* **151**, 351–361.
Doudoroff, M. (1945a). *Fed. Proc., Fed. Am. Soc. Exp. Biol.* **4**, 241–247.
Doudoroff, M. (1945b). *J. Biol. Chem.* **157**, 699–706.
Doudoroff, M. (1955). In "Methods in Enzymology" (S. P. Colowick and N.O. Kaplan, eds.), Vol. 1, pp. 225–231. Academic Press, New York.
Doudoroff, M., Kaplan, N., and Hassid, W. Z. (1943). *J. Biol. Chem.* **148**, 67–75.
Doudoroff, M., Hassid, W. Z., and Barker, H. A. (1944). *Science* **100**, 315–316.
Doudoroff, M., Barker, H. A., and Hassid, W. Z. (1947a). *J. Biol. Chem.* **168**, 725–732.
Doudoroff, M., Hassid, W. Z., and Barker, H. A. (1947b). *J. Biol. Chem.* **168**, 733–746.
Doudoroff, M., Barker, H. A., and Hassid, W. Z. (1947c). *J. Biol. Chem.* **170**, 147–150.
Doudoroff, M., Waime, J. M., and Wolochow, H. (1949). *J. Bacteriol.* **57**, 423–427.
Ficini, M. (1956). *Arch. Studio Fisiopathol. Clin. Ricambio* **20**, 620–636.
Fitting, C., and Doudoroff, M. (1952). *J. Biol. Chem.* **199**, 153–163.
Fleming, S. E., and Grootwassink, J. W. D. (1979). *C.R.C. Crit. Rev. Food. Sci. Nutr.* **12**, 1–28.
Fujimoto, A., and Smith, R. A. (1960). *J. Biol. Chem.* **235**, 44–45.
Gold, A. M., and Osber, M. P. (1971). *Biochem. Biophys. Res. Commun.* **42**, 469–474.
Grazi, E., Trombetta, G., and Morisi, F. (1977). *J. Mol. Catal.* **2**, 453–458.
Hassid, W. Z., and Doudoroff, M. (1950). *Adv. Enzymol.* **10**, 123–143.
Hassid, W. Z., Doudoroff, M., Barker, H. A., and Dore, W. H. (1945). *J. Am. Chem. Soc.* **68**, 1465.
Hassid, W. Z., Doudoroff, M., and Barker, H. A. (1944). *J. Am. Chem. Soc.* **66**, 1416–1419.
Hassid, W. Z. Doudoroff, M., Potter, A. L., and Barker, H. A. (1948). *J. Am. Chem. Soc.* **70**, 306–310.
Hehre, E. J., and Hamilton, D. M. (1946). *J. Biol. Chem.* **166**, 77–78.
Hokse, H. (1983). *Starch* **35**, 101–102.
Hollo, J., Szetsli, J., Laszlo, E., and Vandor, E. (1962). *Staerke* **14**, 53–58.
Hulcher, F. K., and King, K. W. (1958). *J. Bacteriol.* **76**, 571–577.
Imshenetskii, A. A., and Kondrat'eva, T. F. (1980). *Mikrobiologiya* (U.S.S.R) **49**, 98–101.
Kagan, B. O., Latker, S. N., and Zfasman, E. M. (1942). *Biokhimiya* **7**, 93–108.
Kelly, S. J., and Butler, I. G. (1980). *Biotechnol. Bioeng.* **22**, 1501–1507.
Koivistoinen, P., and Hyvonen, L. (1980). "Carbohydrate Sweetness in Foods and Nutrition." Academic Press, London.
Larroque, S., Hammar, L., and Whelan, W. J. (1981). 14th F.E.B.S. Congr. Mar. 29-Apr. 4 Poster.
Ledebew, A., and Dikanowa, A. (1935). *Z. Physiol. Chem.* **231**, 271–272.
Leloir, L. F., and Cardini, C. E. (1953). *J. Am. Chem. Soc.* **75**, 6084.
Leonard, O. A. (1938). *Am. J. Bot.* **25**, 78–83.
Machtelinckx, L. (1983). M. Sc. thesis, University of Ghent.
Manners, D. J., and Taylor, D. C. (1965). *Biochem. J.* **94**, 17P.
Marechal, L. R., and Goldenberger, S. H. (1963). *Biochem. Biophys. Res. Commun.* **13**, 106.
Mieyal, J. J., and Abeles, R. H. (1970). *Enzymes* **7**, 515–532.
Mieyal, J. J., Simon, M., and Abeles, R. H. (1972). *J. Biol. Chem.* **247**, 532–542.
Miklailova, L. G., and Bashkovich, G. A. (1969). *Prikl. Biokhim. Mikrobiol.* **5**, 739 (C. A. **72**, 1970,77402).

Mozza, J. C., Akgerman, A., and Edwards, J. (1975). Carbohydr. Res. **40**, 402–406.
Murao, S., Nagano, H., Ogura, S., and Nishino, T. (1985). Agric. Biol. Chem. **49**, 2113–2118.
Okada, G., and Hehre, E. J. (1974). J. Biol. Chem. **24c**, 126–135.
Oparin, A., and Kursanow, A. (1931). Biochem. Z. **239**, 1–17.
Palleroni, N. J., and Doudoroff, M. (1956). J. Biol. Chem. **219**, 957–962.
Parker, K. J. (1980). Biotechnol. Lett. **2**, 194–198.
Sasaki, T., Tanaki, S., Nakagawa, S., and Kainuma, K. (1983). Biochem. J. **209**, 803–807.
Schmiz, K. L., and John, B. (1983). Arch. Microbiol. **135**, 241–249.
Selinger, Z., and Schramm, M. (1961). J. Biol. Chem. **236**, 2183–2185.
Sih, C. J., Nelson, N. M., and McBee, R. H. (1957). Science **126**, 1116–1117.
Silonova, G. V., and Lisovskaya, N. D. (1975). Metody Sov. Biokhim. 67–69 (C. A. **84**, 31321).
Silverstein, R., Voet, J., Reed, D., and Abeles, R. H. (1967). J. Biol. Chem. **242**, 1338–1346.
Taylor, F., Chen, L., Gong, C. S., and Tsao, G. T. (1982). Biotechnol. Bioeng. **24**, 317–328.
Tsai, L. B., Gong, G. S., and Tsao, G. T. (1980). Abstr. Annu. Meet. Am. Soc. Microbiol., Miami Beach, 11–16 May 146, K120.
Vandamme, E. J. (1976). Chem. Ind. (London) **24**, 1070–1072.
Vandamme, E. J. (1981). J. Chem. Technol. Biotechnol. **31**, 637–659.
Vandamme, E. J. (1983). Enzyme Microb. Technol. **5**, 403–416.
Vandamme, E. J., and Derycke, D. G. (1983). Adv. Appl. Microbiol. **29**, 129–176.
Vandamme, E. J., De Laporte, De Vocht, M., and Van Hoe, L. (1981). In "Mikrobielle Enzymproduktion" (H. Rutloff, ed.), pp. 193–208. Akademie Verlag, Berlin.
Vandamme, E. J., Van Loo, J., Machtelinckx, L., and De Laporte, A. (1984a). Abstr. Int. Biotechnol. Symp., 7th, New Delhi 281–282
Vandamme, E. J., Van Loo, J., Machtelinckx, L., Simkens, E., and De Laporte, A. (1984b). Eur. Congr. Biotechnol. 3rd, Munich **I**, 333–338.
Vandamme, E. J., Van Loo, J., and De Laporte, A. (1987). Biotechnol. Bioeng. **29**, 8–15.
Van Loo, J. (1983). M. Sc. thesis, University of Ghent.
Voet, J. G., and Abeles, R. H. (1970). J. Biol. Chem. **245**, 1020–1031.
Weimberg, R., and Doudoroff, M. (1954) J. Bacteriol. **68**, 381–388.
Whistler, R. L., and Durso, D. F. (1950). J. Am. Chem. Soc. **72**, 677–679.
Wolochow, H., Putnam, E. W., Doudoroff, M., Hassid, W. Z., and Barker, H. A. (1949). J. Biol. Chem. **180**, 1237–1242.

Antitumor Anthracyclines Produced by *Streptomyces peucetius*

A. Grein

Ricerca e Sviluppo di Microbiologia Industriale
Farmitalia Carlo Erba
Milan, Italy

I. Introduction

The anthracyclines represent a class of antibiotics which, because some of them have exhibited extraordinary activity in cancer treatment, became the focus of a major field of research worldwide in the 1970s, although they were discovered about 20 years earlier (Brockmann and Bauer, 1950). Some excellent reviews of this class of compounds with regard to chemical structure (Arcamone et al., 1978, 1981) and cancer therapy properties (Muggia et al., 1982) are available.

This chapter reviews most of the experimental work which has been carried out in our laboratories on the microorganism *Streptomyces peucetius*, which was shown to produce groups of compounds within the anthracycline class endowed with cancer therapeutic potentiality, i.e., daunorubicin and doxorubicin (formerly known as daunomycin and Adriamycin, respectively). The history of their discovery has been reviewed several times in recent years (Arcamone, 1983; White, 1983; White and Stroshane, 1984).

II. Cultural and Morphological Characteristics of *S. peucetius*

The type strain *S. peucetius* (from Peucetia, one of the two ancient regions of the present Apulia, where the actinomycete had been isolated in 1957 from a soil sample) has been described by Grein et al. (1963). This strain had been selected during an antibacterial and antifungal screening program carried out in our laboratories, because it showed activity against gram-positive bacteria and against fungi. The culture, when isolated, showed an impressively heterogeneous population, and in order to correlate the various phenotypes with the different biological activities, the need arose to group those strains showing some macroscopic similarities. This led, at first sight, to a tentative classification of the cultures into seven groups, as shown in Table I, each group characterized by well-defined cultural features.

TABLE I

NATURAL VARIABILITY OF Streptomyces peucetius

Strain group	Color of soluble pigment	Color of substrate mycelium	Aerial mycelium	Spores
I	Orange	Carmine	Scanty, whitish	Blue-green (caesious)
II	Lemon yellow	Yellow-red	Absent	Absent
III	Lemon orange	Yellow	Absent	Absent
IV	Raspberry	Raspberry	Whitish	Absent
V	Absent	Rose	Absent	Absent
VI	Orange	Deep raspberry	Whitish	Absent
VII	Lemon orange–violet	Deep raspberry	Whitish	Absent

A subsequent examination showed that the biological activities were displayed by all the cultures of the different groups, being linked to the various exocellular, but mostly endocellular, pigments formed, and usually present in mixtures, as was later determined.

The aerial mycelium proved unreliable as an identifying characteristic. Its presence was sporadic, and it was usually sterile and took the form of the coremium-like structures described by Grein and Spalla (1962). Sporulation, unfortunately rarely observed, turned out to be one of the most reliable characteristics examined in this species, with respect to color, structure of sporophores, and shape of the spores, except for one mutant, which was identified as belonging to a different genus, i.e., Micromonospora (Grein et al., 1980).

For further examination, priority was given to those groups of cultures showing the highest biological activities, and this led to focusing attention on groups I and VII. The latter group wa soon discarded because it turned out to be composed of cultures producing mainly compounds of the rhodomycin family, which at that time was the best known family among the anthracyclines and relatively easy to identify even with the rather limited analytical facilities available. The main research efforts were therefore concentrated primarily on the cultures constituting group I. In addition to anthracyclines, S. peucetius was found to produce antifungal antibiotics of the polyene type, and two of them, a tetraene and a pentaene, were isolated (Cassinelli and Orezzi, 1963). Moreover, along with these substances, a still unidentified complex mixture of metabolites having antibiotic activity was produced (unpublished data). Some of the cultured strain groups listed in Table I are still under investigation.

III. Mutation and Selection

The wild-type strain was, almost as a rule, an extremely low producer of the anthracyclines, so that as soon as the first antitumor activity data became available, an intensive effort was undertaken with this culture with a twofold aim: first, to improve the yield of the anthracycline mixture, in which daunorubicin appeared as the prominent compound and as mainly responsible for the antitumor activity; second, since a relevant phenotypic variation was observable among the cultures assembled within group I, the need arose for a qualitative analysis of the compounds produced by them, with the hope, of course, of detecting analogs with superior therapeutic properties.

Having set up a reliable working procedure employing various media (Table II), analytical detection methods were continuously revised as more and more sophisticated technologies became available throughout the years. A search program was developed for the screening of strains having possible different and better therapeutic potentialities with respect to daunorubicin.

A mutation-oriented selection process was carried out in parallel with the program designed to increase the yield of daunorubicin (first) adn doxorubicin (Arcamone et al., 1969) (later on). A variety of physical and chemical mutagens applied singly or in combination and known for their different modes of action were employed and the populations surviving treatment were carefully examined.

Individual colonies were scored on the basis of two main characteristics: altered colony morphology and pigmentation. The isolation medium reported in Table II served for selection since it allows a wide variety of colonies to grow and pigment production is particularly pronounced. This medium turned out to be unsuitable for aerial mycelium formation and sporulation, characteristics which, on the other hand, were found less important for a qualitative screen. Surviving populations showed an extremely wide variety of pigmentation ranging from unpigmented colonies to all shades of yellow, orange, red, and violet depending on the type of aglycone formed. Among the unpigmented colonies, some were found to produce the sugar moieties only. The maintainance medium reported in Table II permitted weak sporulation for most of the mutants examined and was therefore also used for morphology studies of conidial structures; it was found that no correlation exists between sporulation capacity and type of product formed by these mutants. The results of this screening program for mutants with qualitative changes of the anthracyclines produced are summarized in Fig. 1 and Table III. In Fig. 1 the mutation-selection

TABLE II

Culture Media Used for S. peucetius

Component	Concentration (%)			
	Isolation medium	Maintenance medium	Seed medium	Production medium
Glucose	3	—	—	6
Sucrose	—	2	3.6	—
Cornsteep liquor	—	1	2.4	—
Brewer's dry yeast	1.5	—	—	3
NaCl	0.1	—	—	0.2
KH_2PO_4	0.05	0.4	—	0.1
$CaCO_3$	0.1	0.4	0.9	0.2
$(NH_4)_2SO_4$	—	0.2	0.24	—
$MgSO_4 \cdot 7H_2O$	0.005	—	—	0.01
$FeSO_4 \cdot 7H_2O$	0.0005	—	—	0.001
$ZnSO_4 \cdot 7H_2O$	0.0005	—	—	0.001
$CnSO_4 \cdot 5H_2O$	0.0005	—	—	0.001
Agar	2	2	—	—
Tap water	—	—	—	—

scheme followed is shown and in Table III the compounds isolated from the various mutants are listed; in Table IV some characteristics of a major mutant, S. peucetius var. aureus, are reported. From this table one can see that mutation brought about not only a new class of compounds but also superior overall production of pigments. Along with mutants producing different anthracycline compounds, mutants completely blocked in the biosynthesis of the decaketide could be isolated with a certain frequency among the survivors of mutagenized populations, and these served advantageously for bioconversion experiments inasmuch as some of them maintained their glycosidation capacities.

IV. Biosynthesis

A ϵ-Rhodomycinone and the corresponding glycoside were reported some years ago by Di Marco and Arcamone (1975) to be present among the metabolites contained in the fermentation beers of the original S. peucetius culture, whereas the presence of aklavinone and its glycoside, aclacinomycin A, have been only recently established in two mutants of S. peucetius, S. peucetius var. castaneus (Grein et al., 1984a) and strain M 104 F.C.E. (Grein et al., 1984b), respectively. These last findings underline once more the unique biosynethetic versatility of S. peucetius

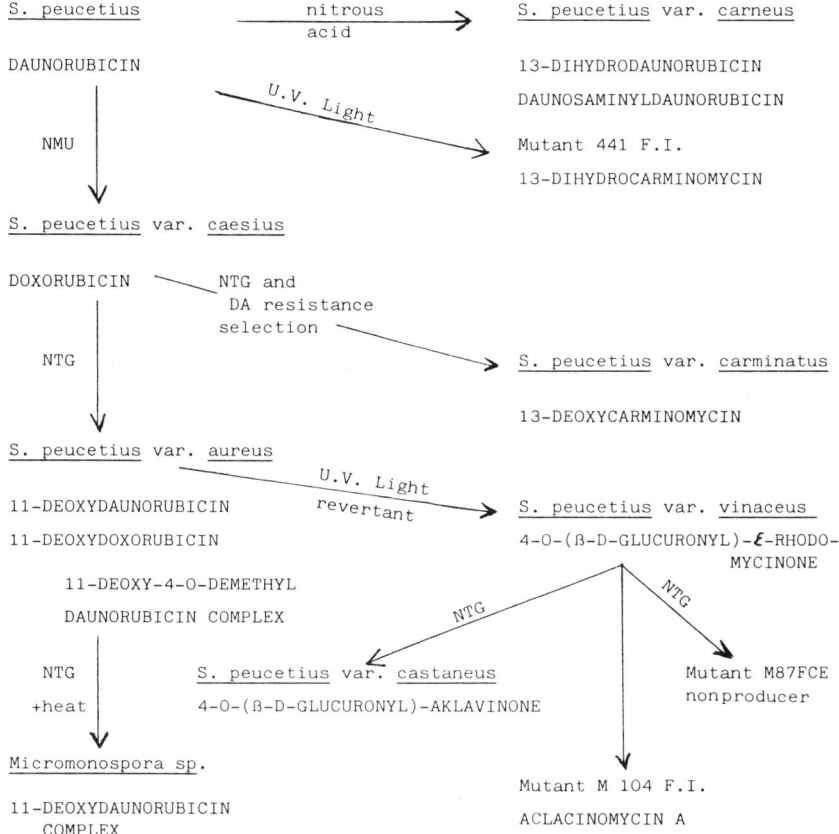

FIG. 1. Mutants of Streptomyces peucetius obtained after different mutagenic treatments. NMU, Nitrosomethylurethane; NTG, N-methyl-N'-nitro-N-nitrosoguanidine; DA, daunorubicin.

as far as both the aglycone and the sugar moieties of this molecule are concerned. In fact, depending upon whether the C-11 position of the aglycone is hydroxylated or not, the cultures are of the red or yellow series respectively, and in turn, within the series, according to the sugar moiety, the rhodomycin or daunomycin group, or alternatively the aclacinomycin or daunomycin analog group, is produced.

In Table V, the four biosynthetic anthracyclines of current therapeutic use are listed and correlated to the known producing species. As can be seen, S. peucetius is the only one capable of producing all four compounds. Among the great variety of strains showing

TABLE III

STRUCTURES OF ANTHRACYCLINES PRODUCED BY S. peucetius AND ITS MUTANTS[a]

Compound[a] (11-hydroxyanthracyclines)	R1	R2	R3	R4	R5	Compound (11-deoxy analogs)	R2
Daunorubicin	COCH$_3$	OH	OCH$_3$	DS	H	11-Deoxydaunorubicin	H
13-Dihydrodaunorubicin	CHOHCH$_3$	OH	OCH$_3$	DS	H	11-Deoxy-13-dihydrodaunorubicin	H
Daunosaminyldaunorubicin	COCH$_3$	OH	OCH$_3$	2DS	H		
13-Deoxydaunorubicin	CH$_2$CH$_3$	OH	OCH$_3$	DS	H	11-Deoxy-13-deoxydaunorubicin	
Doxorubicin (adriamycin)	COCH$_2$OH	OH	OCH$_3$	DS	H	11-Deoxydoxorubicin	
Carminomycin	COCH$_3$	OH	OH	DS	H	11-Deoxycarminomycin	
13-Dihydrocarminomycin	CHOHCH$_3$	OH	OH	DS	H	11-Deoxy-13-dihydrocarminomycin	H
13-Deoxycarminomycin	CH$_2$CH$_3$	OH	OH	DS	H	11-Deoxy-13-deoxycarminomycin	H
Rhodomycin	CH$_2$CH$_3$	OH	OH	TRSC	COOCH$_3$	Aclacinomycin A	H
Rhodomycinone-4-O-(β-D-glucuronyl)	CH$_2$CH$_3$	OH	GA	OH	COOCH$_3$	Aklavinone-4-O-(β-D-glucuronyl)	H
Rhodomycinone-4-O-(β-D-glucuronyl)	—	—	—	OH	H	10-Demethoxycarbonylaklavinone	

L-Daunosamine   L-Rhodosamine   2-Deoxy-L-fucose   L-Cinerulose A   Glucuronic acid

[a] DS, Daunosamine; TRSC, trisaccharide (rhodosamine, 2-deoxyfucose, cinerulose); GA, glucuronic acid.

TABLE IV

COMPARISON OF SOME CHARACTERISTICS OF *S. peucetius* var. *caesius* AND *S. peucetius* var. *aureus*

Characteristic	S. peucetius var. caesius	S. peucetius var. aureus
Color	Red	Yellow
Compounds	Daunorubicin–doxorubicin and related compounds	11-Deoxy-daunorubicin-doxorubicin and related compounds
Total pigments produced	+	+ + +

qualitative differences in anthracyclines, mutants producing mainly ε-rhodomycinone or its glycoside have been found frequently. Through cosynthetic experiments between these strains and strains producing low levels of daunorubicin, as well as by supplying ε-rhodomycinone to nonproducing cultures, it has been possible to establish that ε-rhodomycinone is transformed to daunorubicin (unpublished data), results which are in agreement with those of McGuire et al. (1980), who have demonstrated in a daunorubicin-producing species of *Streptomyces* that ε-rhodomycinone is a precursor of daunorubicin.

That daunorubicin is in turn a precursor of doxorubicin has been demonstrated by labeling experiments in our laboratores (Crespi-Perellino et al, 1982) as well as by Oki et al. (1981) through bioconversion experiments on a doxorubicin-negative mutant of *S. peucetius*. As shown in Table V, where all the known daunorubicin-producing species are listed, *S. peucetius* appears to be the only one possessing the very peculiar enzyme (or enzymes) that convert daunorubicin to doxorubicin.

Glycosidation capability is a major characteristic of *S. peucetius* and probably not fully exploited yet. Although in most cases the anthracyclines isolated from the fermentation beers of this species were found as glycosides, a remarkable versatility in the carbohydrate moieties is observed, ranging from mono- to di- and trisaccharides and glucuronic acid (Cassinelli et al., 1984b), as shown in Figure 2, in addi tion to the various baumycins (White and Stroshane, 1984) which are always present among the fermentation products.

Based upon the results reported above and on the numerous idiotrophs and compounds (Table III) of *S. peucetius* isolated by us, a pathway for biosynthesis of doxorubicin glycoside and its 11-deoxy analogs is proposed as illustrated in the very simplified scheme of Fig. 2, which does not take into account all the possible side and shunt products identified

TABLE V
Organisms Producing Anthracyclines of Current Therapeutic Use

Species	Daunorubicin	Doxorubicin	Carminomycin	Aclacinomycin
Streptomyces peucetius (Grein et al., 1963)	+	+	+	+
Streptomyces coeruleorubidus (Pinnert and Hinet, 1976)	+	−	+	−
Streptomyces galilaeus (Oki et al., 1979)	−	−	−	+
Streptomyces griseus (Mancy and Ninet, 1976)	+	−	−	−
Streptomyces griseoruber (Higashide et al., 1972)	+	−	−	−
Streptomyces viridochromogenes (Liu and Rao, 1974)	+	−	−	−
Streptomyces bifurcus (Mancy et al., 1975)	+	−	−	−
Streptomyces sp. (McGuire et al., 1979)	+	−	−	−
Streptomyces insignis (Tunac et al., 1985)	+	−	−	−
Actinomadura carminata (Gause et al., 1974)	−	−	+	−

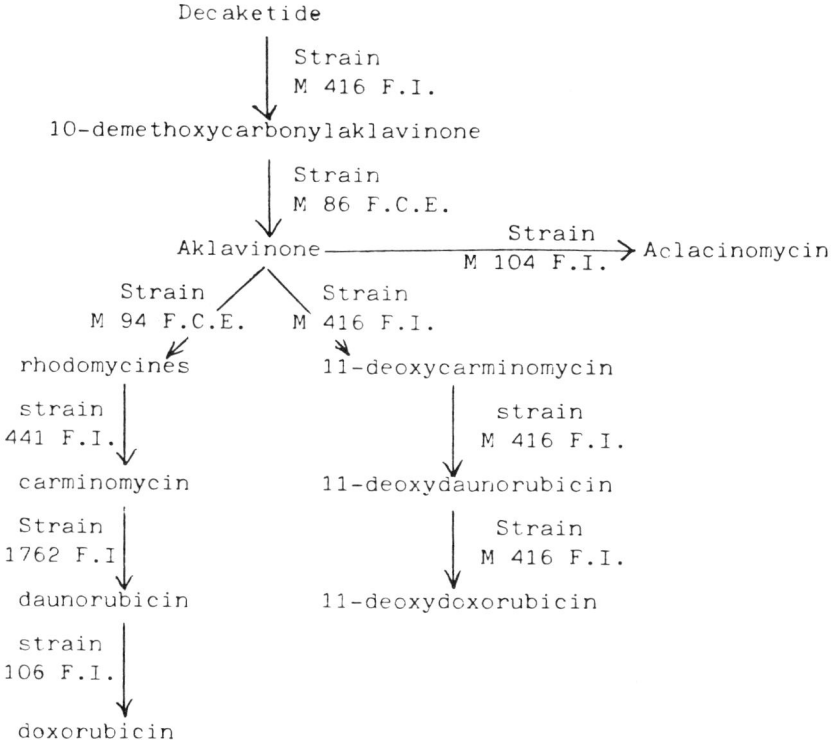

FIG. 2. Proposed scheme for the biosynthesis of anthracyclines produced by S. peucetius. The single arrows between compounds represent an undertermined number of enzymatic steps. This scheme does not necessarily represent a unique pathway of doxorubicin biosynthesis and its analogs, but is consistent with biotransformation data.

in the fermentation beers, such as the ethyl analogs and the dihydroanalogs, as well as higher glycosides and others.

V. Bioconversion

The rather frequent isolation of completely blocked mutants os S. peucetius, present in the populations of all the groups listed in Table I, prompted an investigative program on bioconversion of natural, semisynthetic, and synethetic anthracycline molecules, the latter ones prepared by our chemical laboratories (Penco, 1980).

Some mutants, such as strain M 87 F.I., were able to stereoselectively reduce 13-keto compounds to the corresponding 13-dihydro derivatives

TABLE VI

BIOCONVERSION OF SOME ANTHRACYCLINES BY A BLOCKED MUTANT OF S. peucetius[a]

Substrate	Product
Doxorubicin	13-Dihydrodoxorubicin (adriamycinol)
4'-Epidoxorubicin	13-Dihydro-4'-epidoxorubicin (epirubicinol)
4'-Deoxydoxorubicin	13-Dihydro-4'-deoxydoxorubicin (esorubicinol)
4-Demethoxydaunorubicin	13-Dihydro-4-demethoxy-daunorubicin (idarubicinol)

[a]*Streptomyces peucetius* strain M 87 F.C.E.

which were identical to mammalian metabolites, as reported by Cassinelli et al. (1984a). Moreover, some blocked mutants currently under investigation in our laboratories perform very unusual and not yet reported bioconversions of aglycones (manuscript in preparation).

In Table VI some results of our bioconversion studies are reported which indicate the great potential of research leading to the discovery of new, biologically active molecules, as well as being a useful tool for the preparation of mammalian drug metabolites.

VI. Fermentation

The fermentation of daunorubicin and doxorubicin has been thoroughly described by McGuire et al. (1979) and Grein (1981) so that only a few general considerations may be aded. Both of the phenomena demonstrated in S. peucetius, i.e. that ϵ-rhodomycinone is a precursor of daunorubicin and that in turn daunorubicin is a precursor of doxorubicin, contributed much to the understanding of the fermentation process and its improvement. Moreover, Kralovkova et al. (1977) described a positive effect of barbituric acid on anthracycline biosynthesis by *Streptomyces galilaeus*. We were able to confirm these results for daunorubicin production by S. peucetius and were able to isolate from fermentation beers to which sodium barbiturate had been added a new compound resulting from a complex of daunorubicin and its higher glycosides with barbituric acid (Cassinelli et al, 1983). This new anthracycline showed 10–30 times higher activity against P388 leukemia *in vitro* and *in vivo* compared to doxorubicin, and its structural identification is in progress.

The most critical parameters encountered in studying fermentation conditions appeared to be oxygen availability and temperature during the entire biosynthetic process. Knowledge of interactions of these parameters provide for optimal utilization of precursors and control of their development into the desired end products.

VII. Conclusion

Consideration of the research that has been carried out with S. peucetius as reviewed in this article leads to the conclusion that two major problems, one more theoretical, the other pragmatic, deserve further attention. The first concerns the biosynthesis of the anthracyclines, which seems far from being fully elucidated and is still an object of dispute among various research groups. The experience gathered over many years convinced us that the conflicting results that have been reported, and which are mainly concerned with the role of ϵ-rhodomycinone, may be due to the use of different producer organisms and mutants, and this may reflect that there is not a unique sequence of biosynthetic steps leading to daunorubicin. The second problem concerns the strategy to follow in the future for the discovery of new and more active, as well as better tolerated, compounds of this class. The bioconversion approach to biosynthetic, semisynthetic, and synthetic molecules should still be pursued, taking advantage of the great versatility in glycosylation mechanisms observed in S. peucetius. Information to be derived from the mutation-selection approach appears to be have been nearly exhausted, at least from a practical point of view.

The many different S. peucetius mutants isolated throughout the years have yielded a great variety of molecules, among which doxorubicin still appears to be the most active and the end product of the biosynthetic pathway. All the other molecules discussed here have turned out to be precursors of doxorubicin or biosynthetic shunt products, and show a lower biological activity. There is currently interest in utilization of recombinant DNA technology as applied to S. peucetius. The capability now clearly exists to create new gene combinations and to isolate, study, and manipulate those of more interest. This in turn could lead to new and possible more active molecules. This kind of approach is currently in progress in our microbiological research laboratories.

REFERENCES

Arcamone, F. (1978). Top. Antibiot. Chem. 2, 102–239.
Arcamone, F. (1981). Med. Chem. Ser. 17.
Arcamone, F. (1983). Chron. Drug Discover. 2, 171–188
Arcamone, F., Cassinelli, G., Fantini, G., Grein, A., Orezzi, P., Pol, C., and Spalla, C. (1969). Biotechnol. Bioeng. 9, 1101–1110.
Brockmann, H., and Bauer, K. (1950) Naturwissenschaften 37, 492–493.
Cassinelli, G., and Orezzi, P. (1963). Microbiol. 11, 167–174.

Cassinelli, G., Grein, A., Merli, S., and Rivola, G. (1983). *Antibiot. Belg.* **BE** 896, 190.
Cassinelli, G., Grein, A., Merli, S., Penco., S., Rivola, G., Vigevani, A., Zini, P., and Arcamone, F. (1984a). *Gazz. Chim. Ital.* **114**, 185–188.
Cassinelli, G., Grein, A., Merli, S., and Rivola, G. (1984b). *Glucuronide. Belg.* **BE** 898.837.
Cassinelli, G., Forenza, S., Rivola, G., Arcamone, F., Grein, A., Merli, S., and Casazza, A.M. (1985). *J. Nat. Prod.* **48**, 435–439.
Crespi-Perellino, N., Grein, A., Merli, S., Minghetti, A., and Spalla, C. (1982).*Experientia* **38**, 1455–1456.
Di Marco, A., and Arcamone, F. (1975). *Arzneim. Forsch.* **25**, 368–375.
Di Marco, A., Canevazzi, G., Grein, A., Orezzi, P., and Gaetani, M. (1964). U.S. Patent 4.012.284.
Gause, G. F., Brazhnikova, M. G., and Shoring, V.A. (1974). *Cancer Chemother. Rep.* **58**, 255–256.
Grein, A. (1981). *Proc. Biochem.* **Oct./Nov.**, 34–46.
Grein, A., and Spalla, C. (1962). *Microbiol.* **10**, 175–184.
Grein, A., Spalla, C., Di Marco, A., and Canevazzi, G. (1963). *Microbiol.* **11**, 109–118.
Grein, A., Merli, S., and Spalla, C. (1980). *J. Antibiot.* **33**, 1462–1467.
Grein, A., Merli, S., Rivola, G., and Cassinelli, G. (1984a). *Eur. Conf. Adv. Antitumor Agents, 1st, Milan May 30th–June 1st.*
Grein, A., Merli, S., and Spalla, C. (1984b). *Conv. Soc. Ital. Microbiol. Gen. Biotech. Microb., 3rd, Orvieto Oct. 11#13.*
Grein, A., Merli, S., and Spalla, C. (1985). *Abstr. Int. Symp. Biol. Actinomycetes, 6th, Aug. 26–30, Debrecen* M 85.
Higashide, E., Hasegawa, T., Shibata, M., and Mizumo, K. (1972). U. S. Patent 3.655.878.
Kralovkova, E., Blumauerova, M., and Vanek, Z. (1977). *Folia Microbiol.* **22**, 182–188.
Liu, W. C., and Rao, K. U. (1974). U. S. Patent 3852425.
McGuire, J. C., Hamilton, B. K., and White, R. J. (1979). *Proc. Biochem.* **14**, 2–5.
McGuire, J. C., Thomas, M. C., Stroshane, R. M., Hamilton, B. K., and White, R. J. (1980). *Antimicrob. Agents Chemother.* **18**, 454–464.
Mancy, D., and Ninet, L. (1976). U. S. Patent 3997661.
Mancy, D., Florent, J., and Preud'Homme, J. (1975). U. S. Patent 3855010.
Muggia, F. M., Young, C. W., and Carter, S. K. (1982). *Proc. Int. Symp. Anthracycline Antibiot. Cancer Ther. New York, 16–19 Sept..*
Oki, T., Kitamura, I., Yoshimoto, A., Matsuzama, Y., Shibamoi, N., Ogasawara, T., Inui, T., Takamatsu, A., Takeuchi, T., Masuca, T., Hamada, M., Suda, H., Ishizuka, M., Sawa, T., and Umezawa, H. (1979). *J. Antibiot.* **32**, 791–819.
Oki, T., Takatsuri, Y., Tobe, H., Yoshimoto, A., Takeuchi, T., and Umezawa, H. (1981). *J. Antibiot.* **33**, 1229–1231.
Penco, S. (1980). *Proc. Biochem. June-July* 12–16.
Pinnert, S., and Ninet, L. (1976). U. S. Patent 3989598.
Tunac, J. B., Graham, B. D., Dobson, W. E., and Lenzini, M. D. (1985). *Appl. Environ. Microbiol.* **49**, 265–268.
White, R. J. (1983). Anthracyclines. In "Biochemistry and Genetic Regulation of Commercially Important Antibiotics" (L. C. Vining, ed.), pp. 277–291. Addison-Wesley, Reading, Massachusetts.
White, R. J., and Stroshane, R. M. (1984). *In* "Biotechnology of Industrial Antibiotics" (E. J. Vandamme, ed.), pp. 569–594. Dekker, New York.

INDEX

A

Acetate, in 2,3-butanediol bacterial production
 induction of enzymes, 100
 medium supplement effect, 110, 115–116
Acetobacterium woodii, requirement for nickel, 26
Acetoin reductase
 pH effect, 102
 pyruvate conversion to 2,3-butanediol, 99, 100
 pathways of, 101–102
 stereospecificity, 101
Acetolactate decarboxylase
 pyruvate conversion to 2,3-butanediol, 99, 100, 101
Acetolactate synthase
 induction by acetate, 100
 in mixed acid–2,3-butanediol pathway, 99
 pH optimum, 100
Aclacinomycin, production by Streptomyces species, 208 (table)
Aeration, effect on 2,3-butanediol production
 glucose as substrate and, 118–122
 measurements
  air volume/culture unit volume/minute (VVM), 117–120
  dissolved oxygen tension (DOT), 117–120
  maximum oxygen utilization rate, 118
  rate of oxygen transfer to culture, 117–118
 shake flask cultures, 118
 xylose as substrate and, 120–122
Aeromonas hydrophila
 2,3-butanediol generation, 94
  stereoisomers of, 93 (table)
 diol–hydrogen fermentation, 102, 103
 glucose dissimilation, 91, 104 (tables)
Algae
 corrosion-inducing species, 10
 identification in biofilm on metals, 14–15
 laminaribiose phosphorylase, 165, 166
 trehalose phosphorylase, 165, 166
Aluminum, microbial corrosion
 Cladosporium and, 24–25
  pitting potential decrease by C. resinae, 25
 Pseudomonas and, 24–25
 SRB and, 24
Amines, microbial, as corrosive agents, 19
Ammonia, microbial, in copper and brass corrosion, 19, 23
Animal wastes, bioconversion to methane, 48
 digestion profitability
  beef feedlot, 57, 59–63
  dairy cattle, 53–59
  poultry, 64–66, 67
  swine, 64
 input baseline variables, 39–40
 process parameters, 49, 50
Anthracyclines
 antitumor, production by Streptomyces species, 208 (table)
 bioconversion by blocked S. peucetius mutants, 209–210

216   INDEX

production by S. peucetius
  biosynthesis, 204–205, 207, 209
    scheme of, 209
  mutagen-induced mutants and, 203–205
  strain heterogeneity and, 202
  structure, 206 (table)
Arsenate, effects on sucrose phosphorylase
  good glucosyl acceptor for, 177
  inhibition of, 182
Aspergillus, corrosion-inducing species, 10
Azotobacter xylinum, sucrose phosphorylase activity, 166

B

Bacillus polymyxa (B. polymyxa)
  2,3-butanediol production, 89, 92–93
    from fed wheat, 143
    from starchy feedstock, 157
    stereoisomers of, 93 (table)
  diol–hydrogen fermentation, 102–103
  glucose dissimilation, 91, 104 (tables)
Bacillus subtilis
  2,3-butanediol generation, 94
    stereoisomers of, 93 (table)
  diol–glycerol fermentation, 103
  glucose dissimilation, 91, 104 (tables)
Bacteria
  butanediol-producing
    classification, 90
    culture technique, 134 (table)
      continuous cultivation, 133, 135
      double-fed batch, 133
      immobilized, 135
      simple batch, 133
      two-stage system, 135
    as facultative anaerobes, 116
    fermentation classes, 102–103
    glucose dissimilation
      aeration effect, 91 (table)
      anaerobic products, 104 (table)
    species, 92–94
  corrosion-associated
    damage of
      aluminum, 24–25
      copper, 22–24
      ferrous alloys, 21–22
      lead and zinc, 25–26
    growth prevention and control, 27–30
    identification technique, 14–15
    "iron," aerobic, 8–9
    mechanism of action, 11–24
    miscellaneous, 9
    sulfate-reducing (SRB), anaerobic, 1–3, 7–8
    sulfur-oxidizing, aerobic, 8
Beef feedlot waste, digestion profitability
  economic analysis, 59–63
  herd size and, 57, 59, 60
Benyl viologen, as electron acceptor for SRB, 12, 15
1,3-Butadiene, production from 2,3-butanediol, 152
2,3-Butanediol, bacterial production
  conversion from pyruvate, 96–103; see also Pyruvate
  enzymes, 100–102
  environmental factor effects
    aeration, 116–122
    carbon sources, 104–105
      anaerobic conditions, 106 (table)
    inoculum acclimatization, 123–124
    medium supplements
      acetate, 110, 115–116
      trace metals, 115
      urea, 113–114
      yeast extract, 112–114
    pH, 107–108
    product concentration, 111–112
    substrate concentration
      glucose, 108–109
      oxygen supply and, 108–111
      xylose, 109–111
    temperature, 105, 107
    water activity, 121, 123
  fermentation classes, 102–103
  metabolic functions, 103
  microbiology, 90–95; see also specific bacteria
  potential substrates, 125 (table)
    agricultural residues, 132
    food industry wastes, 126–127
    molasses, 127–128
    sugar beet pulp, 128
    waste sulfite liquor, 124, 126
    wood hydrolysates, 128–132
  process design
    barley as substrate for B. polymyxa, 151

INDEX 217

molasses as substrate, 139–146; see also Molasses
wheat as substrate, 143–144, 146–151; see also Wheat
pyruvate formation and, 95–98
reactor operation mode, 134 (table)
recovery from fermentation broths, 135–139
  countercurrent stream stripping, 137, 139
  solvent extraction
    after adsorption on active carbon, 136–137
    butanediol diacetate-based, 136, 139
    with n-butanol, 136, 138
    preliminary treatment, flowsheet for, 136, 137
2,3-Butanediol, conversion to 1,3-butadiene
diol acetylation, 152
pilot plant pyrolysis unit, 152
n-Butanol, as solvent in 2,3-butanediol recovery, 136, 138

C

Carbon dioxide, production from biomass, 55–56
Carminomycin, production by *Streptomyces* species, 208 (table)
Cattle, waste digestion profitability
carbon dioxide credits and, 55–56
confinement cost and, 54–55
fertilizer credits and, 53–54
fuel escalation rate and, 56, 57
herd size and, 53
interest rate and, 56, 58, 59
Cellobiose phosphorylase
detection in microorganisms, 166
reaction catalyzed by, 164
Cellulose
hydrolysate, conversion to 2,3-butanediol, 132
preparation from aspen wood chips, 131–132
Cerostomella, steel and aluminum corrosion, 10
Cladosporium, concentration cell formation, 14
C. resinae

aluminum corrosion, 9–10
pitting potential decrease by, 25
carboxylic acid as corrosive agent, 17
magnesium corrosion, 26
protective film disruption on metals, 19–20
Clostridium pasteurianum, sucrose phosphorylase activity, 166
Copper, microbial corrosion
ammonia-induced, 23
in polluted seawater, 23
Pseudomonas role, 24
in underground pipes, *Thiobacillus* and SRB role, 23
Corrosion, see Metals, microbial corrosion

D

Daunorubicin
antitumor activity, 201
biosynthesis by *S. peucetius*, 201, 205, 208
conversion to doxorubicin, 207, 210
fermentation, 210
ϵ-rhodomycinone conversion to, 207, 210
production by *Streptomyces* species, 208 (table)
Desulfomaculum
anaerobic corrosion induction, 8
spore-forming, five species, 8
Desulfovibrio
anaerobic corrosion induction, 8
benyl viologen as electron acceptor, 12, 15
non-spore-forming, seven species, 8
volatile phosphorus compound as corrosive agent, 15
Diethyl ether, as solvent in 2,3-butanediol recovery, 136
Disaccharide phosphorylases, 164–165 (table)
DOT, see Aeration, dissolved oxygen tension
Doxorubicin
antitumor activity, 201
production by *S. peucetius*, 201, 205, 208
bioconversion by blocked mutants, 210

daunorubicin conversion to, 207, 210
fermentation, 210

E

Electricity, generation from digester-produced methane, 37, 38, 50, 53, 57, 84

F

Ferrous alloys, microbial corrosion
  cast iron, SRB effects, 21
    ductile, 21–22
    grey, 21–22
  mild steel, SRB effects, 21
  stainless steel, *Gallionella* and SRB effects, 22
Ferrous sulfide, as depolarizer in corrosion, 13–14
Fertilizers, organic
  plant design for digester solid use, 50, 52
  production by anaerobic digestion, 37, 38, 53, 54, 57, 84
Fructose
  effects on sucrose phosphorylase
    good acceptor for, 177
    inhibition of, 182
  metabolism by *L. mesenteroides*, scheme, 171
  production from
    corn syrups, 195–196
    sucrose by sucrose phosphorylase, 196
  properties, 195
Fungi, corrosion-associated, 9–10
  identification technique, 14–15
  organic acids as corrosive agents, 17

G

Gallionella
  aerobic corrosion induction, 8–9
  stainless steel damage, 9, 22
  biofilm on immersed metals, 14
Gelatin, sucrose phosphorylase immobilization, 189
Glucose
  dissimilation by diol-producing bacteria
    aeration effect, 91 (table), 118–122
    anaerobic products, 104 (table)
    2,3-butanediol production, 104–106
    nutrient supplements and, 112–114
    substrate concentration and, 108–109
    pyruvate formation, 95–96
    stimulation by acetate, 100
  effects on sucrose phosphorylase
    complex formation and proteolysis, 184, 186
    inhibition, 181, 185
  metabolism by *L. mesenteroides*, scheme, 171
Glucose-1-phosphate
  as glucosyl donor for sucrose phosphorylase, 177, 182, 187
  potential applications, 196–197
  production from sucrose by sucrose phosphorylase, 196–197
    ATP-independent, 199
Glutaraldehyde, in sucrose phosphorylase immobilization, 189, 191, 192

H

Hemicellulose
  hydrolysis, acid or enzymatic, 129–132
  preparation from aspen chips
    flowsheets of procedures, 130–131
    hydrolysis, acidic or enzymatic, 129, 131
    steam explosion, 129
  sugars from hydrolysate, addition to *K. pneumoniae* culture media, 128–129
  2,3-butanediol and other products recovery, 129, 131–132
Hydrogen, SRB corrosive action and, 19
Hydrogenase, SRB-induced cathodic depolarization and, 11–13
Hydrogen sulfate, as SRB corrosive agent, 18
Hydrogen sulfide, as depolarizer in corrosion, 14

K

Klebsiella pneumoniae (*K. pneumoniae*)
  2,3-butanediol generation, 89, 92
  aeration effect

INDEX 219

on glucose as substrate, 118–122
on xylose as substrate, 120–122
from agricultural residues, 132
anaerobic, from various monosaccharides, 106
from citrus canning-plant press juice, 126
from lactose, 127
stereoisomer of, 91 (table)
from sugar beet molasses, 140
from sugar beet pulp, 128
from wood hydrolysate, aspen, 128–129
  cellulose fraction, 132
  hemicellulose fraction, 129, 131
diol–hydrogen fermentation, 102–103
glucose dissimilation, 91, 104 (tables)
growth, water activity and, 121, 123

L

Lactate dehydrogenase, in mixed acid–2,3-butanediol pathway, 99
Lactose, conversion to 2,3-butanediol by K. pneumoniae, 127
Laminaribiose phosphorylase
  detection in algae, 166
  reaction catalyzed by, 165
Lead, microbial corrosion, 25, 26
Leuconostoc mesenteroides (*L. mesenteroides*)
  fructose metabolism, scheme, 171
  glucose metabolism, scheme, 171
  sucrose metabolism, scheme, 171
  sucrose phosphorylase
    discovery, 163, 166
    fermentation
      medium for, 167–169
      pH and, 171–172
      profile, 169–170
      temperature and, 173–174
    immobilization procedures, 189, 191–192
    pH effect, 180

M

Magnesium, microbial corrosion, 26
Maltose phosphorylase
  detection in *Neisseria* spp., 166

reaction catalyzed by, 164
Metals, microbial corrosion
  cathodic depolarization by SRB
    alternative mechanisms
      FeS as depolarizer, 13–14
      gaseous H_2S as depolarizer, 14
    classical theory, 2, 11–13
  corrosion cell formation
    crevice, 6
    oxygen concentration, 6, 7
  corrosion inhibitor breakdown, 20
  corrosive metabolic products
    acids
      organic, 16–17
      sulfuric, 16–17
    ammonia and amines, 19
    hydrogen, 19
    sulfur and sulfuric compounds, 18
    volatile phosphorus compound, 15–16
  economic aspects, 3–4
  electrochemical process, 5–6
  history, 2–3
  immerse metals, colonization
    biofouling, 14–15
    tubercle formation, 15, 16
  microorganisms associated with
    algae, 10
    bacteria, 7–9
    fungi, 9–10
  multiple mechanisms of, 20–21
  prevention and control
    biocides and biostats, 29
    cathodic protection, 28–29
    environment selection, 27
    plastic tape coating, 28
    protection against SRB, 27–28
  protective film disruption, 19–20
Methane, economics of production from biomass
  capital costs, 49
  computer model, 69, 71
  computer program, overview, 38, 40–48
  digester system, 49–51
    schematic design, 51
  digestion process optimization, 80–84
  digestion profitability
    beef feedlot waste, 57, 59–63
    cattle waste, 53–59
    poultry waste, 64–66, 67
    swine waste, 64
    water hyacinth, 66, 68–69, 70, 84

electricity production from methane and, 37–38, 50, 53, 57, 84
fertilizer plant design and, 50, 52
gas production calculation, 71–72
reaction rate constant and, 75, 76
refractory solid concentration and, 72–75
retention time and, 72, 73
total solid concentration and, 72, 74
input baseline variables, 39–40
organic fertilizer production and, 37–38, 53–54, 57, 84
process parameters, 49, 50
substrates for anaerobic digestion, 48
unit gas cost, effects of
reaction rate constant
with fertilizer plant, 78–79, 80
without fertilizer plant, 77–78, 79
refractory volatile solid concentration
with fertilizer plant, 75–77, 78
without fertilizer plant, 75, 77
refractory volatile solid concentration and reaction rate constant
with fertilizer plant, 80, 82
without fertilizer plant, 79–80, 81
Methanobacterium thermoautotrophicum
nickel-, cobalt-, and molybdenum-dependent strain, 26
N-Methyl-N'-nitro-N-nitrosoguanidine (NTG)
S. peucetius mutagenesis induction, 205
Molasses, 2,3-butanediol production
commercial, process design, 139–146
capital costs, 143, 145, 154–155
estimated costs, 143, 145–146
fermentation, 140–141, 143
flowsheet of plant section, 141
inoculation with K. pneumoniae, 140
production costs, 143, 146, 155–157
product recovery, 140
flowsheet of plant section, 142
experimental, 127–128

N

Nickel, microbial corrosion, 26
Nitrosomethylurethane (NMU)
S. peucetius mutagenesis induction, 205

O

Organic acids, as fungal corrosive agents, 17
Organic chemicals, produced from biomass
cost and production in United States, 154 (table)
demand in United States, 152–153

P

Penicillium, corrosion-inducing species, 10
P. expansum, 2,3-butanediol production, 94–95
Pentitols, metabolism by K. pneumoniae, 95, 98
Pentoses, see also Xylose
metabolism by K. pneumoniae, 95–96, 98
Phosphate, inorganic
determination, enzymatic assay, 197
effects on sucrose phosphorylase
good glucosyl acceptor for, 177
inhibition of, 182
Phosphorus compound, volatile, as corrosive agent, 15–16
Poultry, waste digestion profitability
conversion to methane and other vendable products, 64–65
flock size effects on
mesophilic conversion, 65
digester type and, 65, 67
thermophilic conversion, 65–66
Pseudomonas
biofilm on immersed metals, 14
copper alloy corrosion, 24
corrosion-inducing species, 9
protective film disruption on metals, 19
P. putrefaciens, sucrose phosphorylase activity, 163, 166
P. saccharophila, sucrose phosphorylase
discovery, 163
fermentation media for, 168
immobilization procedures, 188–189
purification, 174–175
Pyruvate
formation from
glucose, 95–96

INDEX

pentoses and pentitols, 95–96, 98
  xylose, 95, 97
mixed acid–2,3-butanediol pathway,
  96–103
Pyruvate dehydrogenase multienzyme
  complex
  aerobic, acetyl-CoA production, 98

R

Rhizopus nigrans, 2,3-butanediol production, 94–95
ϵ-Rhodomycinone, conversion to daunorubicin by *S. peucetius*, 207, 210

S

Serrata marcescens
  2,3-butanediol generation, 94
    stereoisomers of, 93 (table)
  diol–formate fermentation, 103
  glucose dissimilation, 91, 104 (tables)
SRB, see Sulfate-reducing bacteria
Starch, microbial conversion to sucrose
  enzymes, 193–195
  by starch and sucrose phosphorylases
    in batch reactor, 193
Starch phosphorylase
  glucose-1-phosphate production, 196
  sucrose production, 193–194
    combined action with sucrose phosphorylase, 193
Streptomyces peucetius
  anthracycline production
    heterogeneity in strain groups, 202
    mutation-oriented selection and,
      203–204
      high-yielding mutants, 205
      var. *aureus*, characteristics, 204,
        205, 207
    blocked mutants, anthracycline bioconversion, 210
  culture media, 204 (table)
  strain groups, morphology, 201–202
Streptomyces species, antitumor anthracycline production, 208 (table)
Sucrose
  conversion by sucrose phosphorylase to
    fructose, 163–164, 196

  glucose-1-phosphate, 163–164,
    196–197
    determination, enzymatic assay, 197
  glucose-labeled, synthesis by sucrose
    phosphorylase, 197–198
  as glucosyl donor for sucrose phosphorylase, 177, 182, 187
  metabolism by *L. mesenteroides*,
    scheme, 171
  microbial production from starch, 193
    by combined action of starch phosphorylase and sucrose phosphorylase in batch reactor, 193
  enzymes, 193–195
  worldwide consumption increase,
    192–193
Sucrose phosphorylase
  activity changes by
    pH, 176, 180
      buffer systems and, 179 (table)
    temperature, 175–176, 177
  applications
    fructose production from sucrose,
      196
    glucose-labeled sucrose synthesis,
      197–198
    glucose-1-phosphate production from
      sucrose, 196–197
    high energy P-bond formation without ATP, 199
    inorganic phosphate enzymatic determination, 197
    novel disaccharide production,
      198–199
    sucrose production from starch,
      193–194
  discovery in microorganisms, 163, 166
  fermentation
    media, composition, 167–169
    pattern, 169–170
      pH and, 171–172
      temperature and, 172, 173
  glucose–enzyme complex
    formation, 184, 186
    proteolysis, 186
  immobilization, 188–192
    pH effect on activity, 189, 190
    procedures, 189, 191–192
    productivity in continuous column
      reactors, 190–191
  inhibition by

glucose, 181, 185
glucosyl acceptors, 182, 186
  dissociation constants, 186 (table)
kinetic constants with various donors and acceptors, 187 (table)
liberation from bacterial cells, 173–174
mode of action, 186, 188
physicochemical properties, 178 (table)
purification, 174–175
substrate specificity
  for glucosyl acceptors, 183, 184 (tables)
    good, 177, 180, 181
    poor, 177
  for glucosyl donors, 177, 182 (table)
Sugar beet
  molasses, 2,3-butanediol production
    commercial, 136–146
    experimental, 127–128
  pulp, conversion to 2,3-butanediol by K. pneumoniae, 128
Sulfate-reducing bacteria (SRB)
  cathodic depolarization of metals
    FeS as depolarizer, 13–14
    gaseous H_2S as depolarizer, 14
    hydrogenase system, 12–13
  corrosion of
    ferrous alloys, 21–22
    zinc, 25–26
  corrosive agents
    hydrogen sulfide, 18
    volatile phosphorus compound, 15
  growth prevention by
    alkaline environment, 28
    anaerobic condition avoidance, 27
    biocides and biostats, 29
    protective film disruption on metals, 13, 19
Sulfuric acid, sulfur-oxidizing bacteria
  as corrosive agent, 16–17, 23
    effects on underground copper pipes, 23
Swine, waste digestion profitability, 64
  herd size effect, 64 (table)

T

Thiobacillus
  aerobic corrosion induction, 8
  iron oxidation, 8
  sulfuric acid as corrosive agent, 8, 17, 23
  underground copper pipe damage, 23
Trace metals, medium supplements
  2,3-butanediol bacterial production and, 115
Trehalose phosphorylase
  detection in *Euglena gracilis*, 166
  reaction catalyzed by, 165

U

Urea, medium supplement
  2,3-butanediol bacterial production and, 113–114

V

VVM, see Aeration, air volume/culture unit volume/minute

W

Wastes
  animal, see Animal wastes
  sulfite liquor, butanediol production from, 124–126
Water activity, 2,3-butanediol-producing bacteria and, 121, 123
Water hyacinth
  bioconversion to methane, 48
    computer program for, 44–47
    input baseline variables, 39–40
    process parameters, 49, 50
  digestion profitability
    electricity generation and, 84
    fertilizer production and, 84
    pond size and, 68–69, 70
    technical and cost parameters, 66, 68
  production in water effluents, 52–53
Wheat, 2,3-butanediol commercial production
  B. polymyxa inoculation, 144
  estimated cost, 151
  fermentation, 146
  flowsheet of operations, 147
  mass flow rates, 150
  plant design, 148–151
  recovery, 148

INDEX

Wood hydrolysates, butanediol production from, 128–132; see also Hemicellulose

X

Xylose
  2,3-butanediol bacterial production, 104–106
  aeration of media and, 120–122
  nutrient supplements and, 112–114
  substrate concentration and, 109–111
  conversion to pyruvate, 95, 97

Y

Yeast extract, medium supplement
  2,3-butanediol bacterial production and, 112–114

Z

Zinc, microbial corrosion
  SRB role, 25–26

RAYMOND H. FOGLER LIBRARY
DATE DUE

SUBJECT TO